MODERN CONTROL SYSTEMS

ANALYSIS AND DESIGN

USING MATLAB AND SIMULINK

Robert H. Bishop

The University of Texas at Austin

A companion to

MODERN CONTROL SYSTEMS

SEVENTH EDITION

Richard C. Dorf

Robert H. Bishop

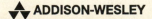 **ADDISON-WESLEY**

An imprint of Addison Wesley Longman, Inc.

Menlo Park, California · Reading, Massachusetts · Harlow, England
Berkeley, California · Don Mills, Ontario · Sydney · Bonn · Amsterdam · Tokyo · Mexico City

Acquisitions Editor: Tim Cox
Assistant Editor: Laura Cheu
Editorial Assistant: Royden Tonomura
Production Editor: Lisa Weber
Manufacturing Coordinator: Janet Weaver
Cover Design and Illustration: Yvo Riezebos
Text Designer: Margaret Ong Tsao and Arthur Ogawa
Copyeditor: Barbara Conway
Proofreader: Joe Ruddick

Camera-ready copy for this book was prepared by the author using LATEX 2_ε.

Many of the designations used by manufacturers and sellers to distinguish their products are claimed as trademarks. Where those designations appear in this book, and Addison Wesley Longman was aware of a trademark claim, the designations have been printed in initial caps or all caps.

MATLAB and SIMULINK are registered trademarks of The MathWorks, Inc.

Instructional Material Disclaimer

The programs presented in this book have been included for their instructional value. They have been tested with care but are not guaranteed for any particular purpose. Neither the publisher nor the author offer any warranties or representations, nor do they accept any liabilities with respect to the programs.

Library of Congress Cataloging-in-Publication Data

Bishop, Robert H., 1957–
    Modern control systems analysis and design using Matlab and
Simulink / Robert H. Bishop.
        p.   cm.
    Companion text to: Modern control systems. 7th ed. / Richa
    Includes index.
    ISBN 0-201-49846-4
    1. Feedback control systems. 2. MATLAB. 3. SIMULINK
4. Computer-aided design.   I. Title.
    TJ216.B53   1996
    629.8′3′011353042--dc21   96-46978
                                                            CIP

ISBN 0-201-49846-4

1 2 3 4 5 6 7 8 9 10 CRS 01 00 99 98 97

Addison Wesley Longman, Inc.
2725 Sand Hill Road
Menlo Park, California 94025

Dedicated to:

Lynda, Bobby and Joey

In grateful appreciation

C O N T E N T S

Preface   **viii**

CHAPTER 1   **Introduction to Control Systems**   **1**
             A Space Shuttle Example

             1.1   Introduction   2
             1.2   The Design Process   5
             1.3   Simulating a Simple System with SIMULINK   7
             1.4   Summary   16
                       Exercises   16

CHAPTER 2   **Mathematical Models of Systems**   **17**
             Fluid Flow Modeling Example

             2.1   Introduction   17
             2.2   Fluid Flow Modeling   18
             2.3   Mathematical Model and Assumptions   20
             2.4   Differential Equations of Motion   22
             2.5   Solutions to the Equations of Motion   26
             2.6   Summary   33
                       Exercises   33

CHAPTER 3   **State Variable Models**   **34**
             A Space Station Example

             3.1   Introduction   35
             3.2   Two Simple Physical Systems   35
             3.3   Spacecraft Control   43
             3.4   Simplified Nonlinear Model   53
             3.5   Linearization   54
             3.6   Pitch-axis Analysis   56
             3.7   Summary   60
                       Exercises   60

CHAPTER 4   **Feedback Control System Characteristics**   **61**
                Blood Pressure Control Example

        4.1 Introduction   62
        4.2 Error Signal Analysis   62
        4.3 Blood Pressure Control During Anesthesia   71
        4.4 Summary   83
                Exercises   83

CHAPTER 5   **Performance of Feedback Control Systems**   **84**
                Airplane Lateral Dynamics Example

        5.1 Introduction   85
        5.2 Airplane Lateral Dynamics   86
        5.3 Bank Angle Control Design   91
        5.4 Simulation Development   95
        5.5 Summary   102
                Exercises   102

CHAPTER 6   **Stability of Linear Feedback Systems**   **104**
                Robot-controlled Motorcycle Example

        6.1 Introduction   105
        6.2 BIBO Stability   105
        6.3 Robot-controlled Motorcycle   111
        6.4 Stability Analysis   114
        6.5 Disturbance Response   116
        6.6 MATLAB Analysis   117
        6.7 Summary   120
                Exercises   120

CHAPTER 7   **Root Locus Method**   **121**
                Automobile Velocity Control Example

        7.1 Introduction   121
        7.2 Sketching a Root Locus   122
        7.3 PID Controller   128
        7.4 Automobile Velocity Control   130
        7.5 Summary   137
                Exercises   138

CHAPTER 8   **Frequency Response Methods**   **139**
Six-legged Ambler Example

8.1   Introduction   140
8.2   A Simple Physical System   142
8.3   Six-legged Ambler   146
8.4   Controller Selection   147
8.5   Controller Design   148
8.6   Summary   154
        Exercises   155

CHAPTER 9   **Stability in the Frequency Domain**   **157**
Hot Ingot Robot Control Example

9.1   Introduction   158
9.2   Hot Ingot Robot Control   158
9.3   Proportional Controller Design   161
9.4   Nyquist Plot for a System with a Time-delay   163
9.5   Padé Approximation   168
9.6   Other Time-delay Approximations   170
9.7   Nyquist Plot with Padé Approximation   171
9.8   PI Controller Design   173
9.9   Summary   178
        Exercises   179

CHAPTER 10   **Design of Feedback Control Systems**   **180**
Milling Machine Control Example

10.1   Introduction   180
10.2   Lead and Lag Compensators   181
10.3   Milling Machine Control System   181
10.4   Lag Compensator Design   187
10.5   Summary   191
         Exercises   191

CHAPTER 11   **Design of State Variable Feedback Systems**   **192**
              Diesel Electric Locomotive Example

        11.1   Introduction   193
        11.2   Robot Drive Train Dynamics   195
        11.3   More on Controllability   199
        11.4   More on Observability   203
        11.5   Diesel Electric Locomotive Example   204
        11.6   State Feedback Controller Design   206
        11.7   System Simulation with SIMULINK   211
        11.8   Summary   214
                   Exercises   215

CHAPTER 12   **Robust Control Systems**   **216**
              Digital Audio Tape Speed Control Example

        12.1   Introduction   216
        12.2   Uncertain Time-delays   217
        12.3   Digital Audio Tape Example   222
        12.4   PID Controller Design   224
        12.5   Summary   229
                   Exercises   229

CHAPTER 13   **Digital Control Systems**   **230**
              Fly-by-wire Control Surface Example

        13.1   Introduction   230
        13.2   Fly-by-wire Aircraft Control Surface   231
        13.3   Settling Time and Percent Overshoot Specifications   236
        13.4   Controller Design   237
        13.5   Summary   240
                   Exercises   241

APPENDIX A   **Useful Design Formulas**   **242**

              References   **245**

              Index   **249**

MODERN CONTROL SYSTEMS—THE SUPPLEMENT

This supplement is designed to be used as a companion to the main textbook, *Modern Control Systems* by Richard C. Dorf and Robert H. Bishop. The primary objective of this supplement is to strengthen the design emphasis by introducing a selected set of solved design problems. Each chapter focuses on one design problem adapted from *Modern Control Systems*. The problems are selected from a wide range of fields. In addition to the design problems, some chapters also include example problems that illustrate important points and concepts. Rather than focus just on the design issues of specific (and interesting) design problems, this supplement is built around the notion of a design process.

Design is the process of conceiving or inventing the forms, parts, and details of a system to achieve a reasoned purpose. Control system design is only one important example of design. Design is a creative endeavor, so there is not a unique methodology that guarantees a valid design solution. To help organize the design process, we suggest a series of steps leading to the final design. Every chapter addresses at least one step of the design process. In general, the early chapters focus on modeling, design specifications, and identification of important variables to be to controlled. The later chapters focus on controller selection and design and analysis of the controlled system. Of course the design process is inherently iterative, so some steps will be repeated as the design is refined. MATLAB and SIMULINK are valuable tools in the design process because they effectively assist in performing the repetitive steps quickly.

Now that powerful computers and software are available for control system design, some may ask the question, Why don't we program the full higher-order nonlinear equations, ignore all the modeling and analysis techniques, and use the computer to grind out an answer? There are many reasons why this is not a good way to solve an engineering problem. First, we do not get a feel for the problem. For example, suppose the team leader tells us the design specifications have been changed for the problem we are currently working. Which controller gain changes to meet the new design specifications? Do we need to change the controller structure? Without a feel for the problem, we may have few ideas on how to proceed. In this supplement we present the technique of obtaining approximate transfer function models to determine initial controller designs and then relying on MATLAB to fine-tune and analyze the closed-loop control system.

THE DESIGN PROCESS

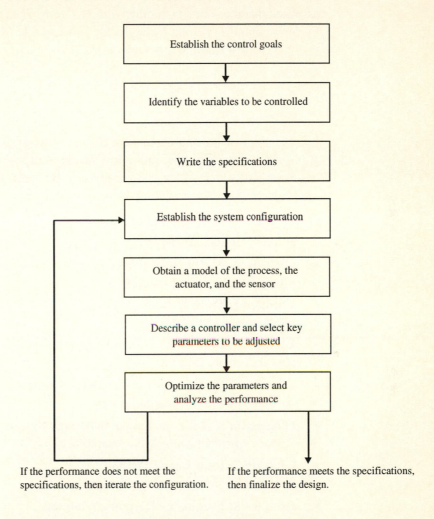

In this manner, we can develop good engineering intuition regarding the design variables and how they affect the system response.

In this book we also use the notion of dominant poles to obtain initial control system designs. The idea is that we design the controller such that the closed-loop system response is dominated by certain poles placed appropriately to meet the design specifications. Again, we can use MATLAB to verify quickly that the design specifications have indeed been satisfied. Each time we use MATLAB in a problem solution in this supplement, we give the associated script. We can use the scripts to verify the results, but more importantly, they can be modified to solve other similar design problems.

To properly utilize this supplement it is essential to have access to *Modern Control Systems*. Many of the problems and examples in Dorf and Bishop are solved here using MATLAB and SIMULINK, but the background information presented in *Modern Control Systems* has not been repeated. For example, it is assumed that the reader is familiar with MATLAB. The main text *Modern Control Systems* contains relevant materials for new users of MATLAB and that material is not presented again in this supplement.

ORGANIZATION

Each chapter of the supplement follows the corresponding chapter in *Modern Control Systems*. To allow the reader to relate the supplement chapters to the main textbook chapters, the chapter titles have remained the same. However, we have added a subtitle indicating the primary design problem of that chapter.

- **Chapter 1: Introduction to Control Systems**
  A Space Shuttle Example

- **Chapter 2: Mathematical Models of Systems**
  Fluid Flow Modeling Example

- **Chapter 3: State Variable Models**
  A Space Station Example

- **Chapter 4: Feedback Control System Characteristics**
  Blood Pressure Control Example

- **Chapter 5: Performance of Feedback Control Systems**
  Airplane Lateral Dynamics Example

- **Chapter 6: Stability of Linear Feedback Systems**
  Robot-controlled Motorcycle Example

- **Chapter 7: Root Locus Method**
  Automobile Velocity Control Example

- **Chapter 8: Frequency Response Methods**
  Six-legged Ambler Example

- **Chapter 9: Stability in the Frequency Domain**
  Hot Ingot Robot Control Example

- **Chapter 10: Design of Feedback Control Systems**
  Milling Machine Control Example

- **Chapter 11: Design of State Variable Feedback Systems**
  Diesel Electric Locomotive Example

- **Chapter 12: Robust Control Systems**
  Digital Audio Tape Speed Control Example

- **Chapter 13: Digital Control Systems**
  Fly-by-wire Control Surface Example

The design problems in each chapter are all adapted from *Modern Control Systems*. In most cases, the problems are end-of-chapter problems revisited. The relationship between the chapter design problems in the supplement and *Modern Control Systems* is shown in the following table.

Supplement Chapter Number	Design Problem	Relationship to *Modern Control Systems*
1	Space Shuttle	P9.9
2	Fluid Flow Modeling	P2.12
3	Space Station Modeling	Section 3.9
4	Blood Pressure Control	AP4.5
5	Airplane Lateral Dynamics	DP5.1
6	Robot-controlled Motorcycle	DP6.6
7	Automobile Velocity Control	DP7.12
8	Six-legged Ambler	DP8.2
9	Hot Ingot Robot Control	DP9.10
10	Milling Machine Control	P10.36
11	Diesel Electric Locomotive	DP11.3
12	Digital Audio Tape Speed Control	DP12.2
13	Fly-by-wire Control Surface	AP13.2

THE SOFTWARE

It is assumed that the readers have access to MATLAB and the *Control System Toolbox*. All of the MATLAB examples in this supplement were developed and tested on a Power Macintosh 7200/90 with MATLAB Version 4.2c and the *Control System Toolbox*. Since it is not possible to verify each example on all the available computer platforms that are compatible with MATLAB, we restrict the computer topics covered in this supplement to those that are platform independent. It will be very helpful to have access to the MATLAB *Users Guide*.

Readers do not need access to SIMULINK to use this supplement effectively. Every design problem is solved using MATLAB, so skip the SIMULINK material if desired. It is clear, however, that SIMULINK provides valuable additional simulation capability; therefore, we introduce it in this supplement for those readers wishing to extend their knowledge base. We used SIMULINK 1.3c in the simulation development. It will be very helpful to also have access to the SIMULINK *Users Guide*.

A set of M-files, the *Modern Control Systems Supplement Toolbox*, have been developed by the author for this supplement. The M-files contain the scripts from each MATLAB example. You can retrieve the M-files from Addison-Wesley at **ftp.aw.com**. Please refer to the Addison-Wesley Computer Science and Engineering Web site at **http://www.aw.com/cseng** or call 1-800-322-1377 if you would like to purchase a copy of *Modern Control Systems*.

ACKNOWLEDGEMENTS

We wish to express appreciation to the following individuals who assisted with the development of the supplement: Peter J. Gorder, Kansas State University; Randall S. Janka, Mercury Computer Systems (CPG); Mariusz Jankowski, University of Southern Maine; L. G. Kraft, University of New Hampshire; Pradeep Misra, Wright State University; Mark L. Nagurka, Marquette University; Hal Tharp, University of Arizona; John Valasek, Western Michigan University; Fred Weber, University of Tennessee, Knoxville; Marcus Benavides and Terry Hill, both undergraduate students at The University of Texas at Austin; Dr. Scott J. Paynter for his contribution to Chapter 3; and Tim Crain for checking the many MATLAB scripts on an IBM-compatible PC. Finally we would like to express appreciation to Lynda Bishop for assisting with the development of the manuscript.

OPEN LINES OF COMMUNICATION

The author and the staff at Addison-Wesley Publishing Company would like to establish an open line of communication with the users of this supplement. We encourage all readers to email Addison-Wesley with comments and suggestions for this and future editions. By doing this, we can keep you informed of any general interest news regarding the supplement and pass along interesting comments from other users.

Keep in touch!

Robert H. Bishop
bishop@zeus.ae.utexas.edu

Addison-Wesley Publishing Company
cse@aw.com

CHAPTER 1

Introduction to Control Systems

A Space Shuttle Example

1.1 Introduction   2
1.2 The Design Process   5
1.3 Simulating a Simple System with SIMULINK   7
1.4 Summary   16
     Exercises   16

P R E V I E W

This chapter provides an introduction to the control design process by describing a general approach to designing and analyzing a feedback control system. We discuss the basic components of control system design in the context of the process of design. We also discuss the notion of the design process used in all of the subsequent chapters. The idea is to emphasize the tight link between the theory and applications and the design process.

We introduce the simulation program SIMULINK in this chapter, which is an extension to MATLAB that enables students and practicing engineers to simulate dynamic systems quickly and effectively. Using block diagram windows, we can create and edit models by manipulating (principally with mouse-driven commands) model components, such as scopes, signal generators, and transfer functions. SIMULINK allows for graphical insight to a problem and helps to develop our intuition.

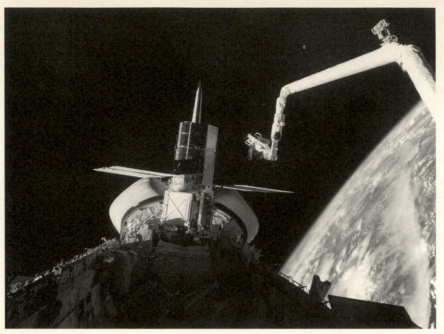

(Photo courtesy of NASA)

FIGURE 1.1
The space shuttle with the deployed Remote Manipulator System (RMS).

1.1   INTRODUCTION

Figure 1.1 shows the space shuttle during on-orbit operations with the robotic arm, known as the Remote Manipulator System (RMS). During shuttle capture and retrieve operations, the vehicle attitude hold control system is used to maneuver the vehicle to a desired attitude and to hold that attitude. The main components of the attitude hold control system are the primary reaction control system, the flight computer hosting the control algorithms, the astronaut piloting the vehicle (when not in attitude hold mode), the rotational hand controller, and the various sensors, such as rate gyros and accelerometers. A block diagram of the shuttle attitude hold control system is shown in Figure 1.2 [1]. Bear in mind that this block diagram is a simplification of the actual implemented control system on board the shuttle.

We can compare the shuttle controller block diagram shown in Figure 1.2 with a block diagram that is more characteristic of the ones found in most undergraduate controls textbooks. Most standard block diagrams have the form shown in the simplified block diagram in Figure 1.3. The following is a typical problem statement associated with the simplified block diagram (see for example P9.9 in *Modern Control Systems*):

Typical Problem
Statement   The key to future exploration and use of space is the reusable earth-to-orbit transport system, popularly known as the space shuttle. The shuttle carries large payloads into space and returns them to earth for reuse [2]. The block diagram of a pitch rate control system is shown in Figure 1.3. The sensor is represented by a gain,

$$H(s) = 0.5,$$

and the vehicle by the transfer function

$$G(s) = \frac{0.3(s + 0.05)(s^2 + 1600)}{(s^2 + 0.05s + 16)(s + 70)}.$$

The controller can be a simple gain or any suitable transfer function. (a) Draw the Bode diagram of the system when $G_c(s) = s$ and determine the stability margin. (b) Draw the Bode diagram of the system when

$$G_c(s) = K_1 + \frac{K_2}{s} \quad \text{and} \quad \frac{K_2}{K_1} = 0.5.$$

The gain K_1 should be selected so that the gain margin is 10 dB.

The space shuttle control system shown in Figure 1.2 has been replaced with

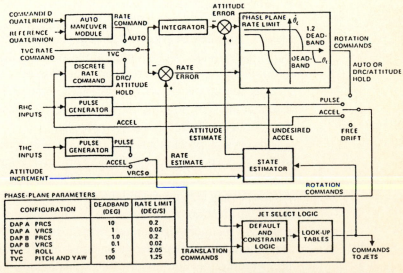

(Copyright © 1982 AIAA—reprinted with permission.)

FIGURE 1.2
The space shuttle attitude hold control system.

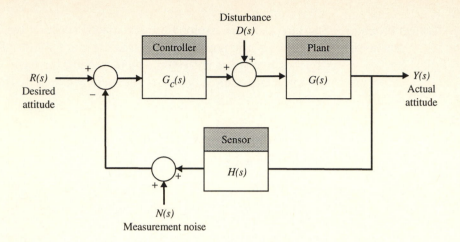

FIGURE 1.3
A simplified representation of the space shuttle attitude hold control system.

a proportional-integral (PI) controller, denoted by $G_c(s)$. The complex space shuttle vehicle dynamics and sensor dynamics are modeled by the transfer functions $G(s)$ and $H(s)$, respectively. Why the difference in complexity of the controllers? Is our approach here merely academic, not real world? What is the connection between the so-called real world and the methods and techniques usually presented in introductory courses in controls? The answers can be found by considering the design process.

The design process is inherently iterative—we must start somewhere! Successful engineers learn to simplify complex systems appropriately for design and analysis purposes. A gap between the complex physical system and the design model is inevitable. **Design gaps** are intrinsic in the progression from the initial concept (Figure 1.3) to the final product (Figure 1.2). We know intuitively that it is easier to improve an initial concept incrementally than to try to create a final design at the start. In other words, engineering design is not a linear process. It is an iterative, nonlinear creative process.

The design process has been dramatically impacted by the advent of powerful and inexpensive computers and effective control design and analysis software. For example, the Boeing 777, which incorporates the most advanced flight avionics of any U.S. commercial aircraft, was almost entirely computer-designed [3, 4]. Verification of final designs in high-fidelity computer simulations is essential. In many applications the certification of the control system in realistic simulations represents a significant cost in terms of money and time. The Boeing 777 test pilots flew about 2400 flights in high-fidelity simulations before the first aircraft was even built.

Another notable example of computer-aided design and analysis is the Mc-Donnell Douglas Delta Clipper experimental vehicle DC-X, which was designed, built, and flown in 24 months. Computer-aided design tools and automated code-generation contributed to an estimated 80% cost savings and 30% time savings [5].

Rapid prototyping is a process by which we can design and analyze systems (including feedback control systems) without actually building tools and die set-ups [6]. The use of computer-aided control design and analysis software (such as MATLAB and SIMULINK) represents a very real step towards bridging the gap between the theory of control system design and the rapid prototyping of controllers in real world situations. It is not the whole story, but it is a step in the right direction.

1.2   THE DESIGN PROCESS

The **design process** is illustrated in Figure 1.4 and is discussed in Section 1.9 of *Modern Control Systems*. The process consists of seven main building blocks, which we can divide into three groups:

1. Establishment of goals and variables to be controlled, and definition of metrics (that is specifications) against which to measure performance

2. System definition and modeling

3. Control system design and integrated system simulation and analysis

The theory topics presented in *Modern Control Systems* correspond to aspects of the design process. We have the following connections between the chapters in *Modern Control Systems* and the design process:

1. Establishment of goals, control variables, and specifications: *Modern Control Systems* Chapters 1, 4, 5, 6, 9, and 13

2. System definition and modeling: *Modern Control Systems* Chapters 2, 3, and 13

3. Control system design, simulation, and analysis: *Modern Control Systems* Chapters 7, 8, 10, 11, 12, and 13

In each chapter of this supplement, we highlight the connection between the design process and the main topics in the companion chapters in *Modern Control Systems*. In particular, Figure 1.4 appears in each of the subsequent chapters with appropriate highlights indicating the specific elements of the design process emphasized in that chapter.

In each chapter we illustrate different aspects of the design process by working examples taken from *Modern Control Systems*. For example, in the study of controls, we are often given a transfer function and told that it represents some

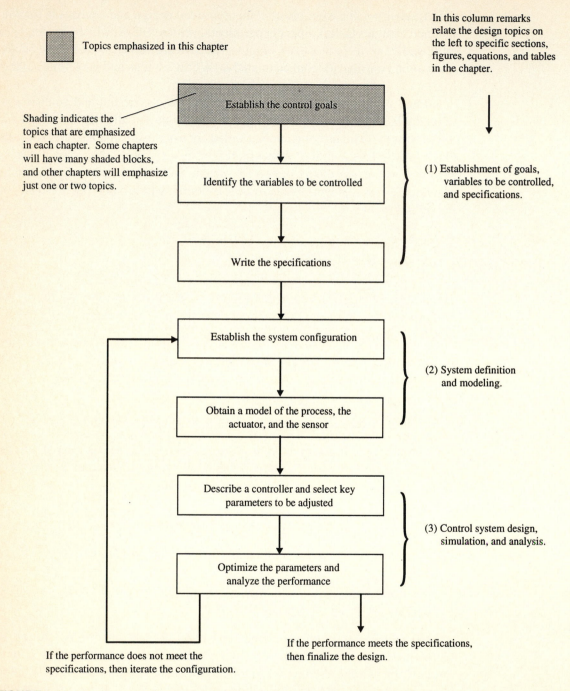

FIGURE 1.4
Elements of the control system design process emphasized in each chapter.

physical system. How the transfer function was obtained is not discussed. We address this situation in this supplement by providing more insight into the modeling aspect: How is the transfer function obtained? What basic assumptions are implied in the model development? How general are the transfer functions?

1.3   SIMULATING A SIMPLE SYSTEM WITH SIMULINK

SIMULINK allows the users of MATLAB to simulate feedback control systems. We view SIMULINK as another tool in our toolbox for designing and analyzing control systems. In this section we walk through a sequence of steps to introduce SIMULINK (see [7, 8]).

After logging into the workstation or personal computer, we will need to start MATLAB. If you do not know how to access MATLAB, you can refer to the *User's Manual* or visit the local help desk.

Once MATLAB is active we should see a screen on our desktop similar to that shown in Figure 1.5. We type simulink at the prompt and then press the Return or Enter key. A window resembling Figure 1.6 appears on the desktop. From here we can open (use File) a pre-existing file or create a new simulation.

In this chapter we will construct a very simple system and initiate a simulation in SIMULINK. In Chapters 5 and 11 we will construct and experiment with significantly more complex systems. Keep in mind that we should not rely solely on the material presented here to get to know SIMULINK and MATLAB. The printed literature that comes with the software is an excellent reference source. Also, for an introduction to MATLAB, refer to Appendix F of *Modern Control Systems* as well as relevant sections throughout *Modern Control Systems*.

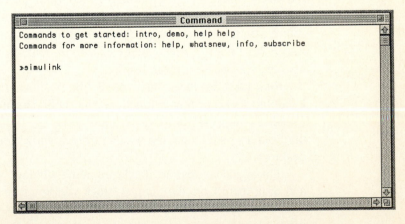

FIGURE 1.5
MATLAB screen from which we initialize the startup of SIMULINK.

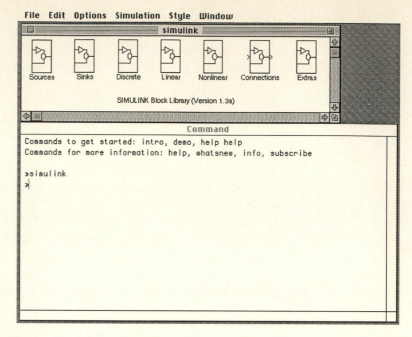

FIGURE 1.6
First window presented by SIMULINK.

Begin by selecting NEW from the File menu, as shown in Figure 1.7. A new system window opens as shown in Figure 1.8. We will build a simple system in this window.

The system window opens in a default position on the desktop. We can move it around and re-size it as desired. This system window is named Untitled when first created. We can rename the window by selecting Save As from the File menu. By saving the system window, we will, in fact, automatically create an M-file. An M-file contains the information the system needs to open the window in future SIMULINK sessions for further editing and simulation. We should choose SAVE from the File menu after each subsequent session so that all changes and updates are saved for future work.

SIMULINK has the standard block library shown in Figure 1.6. It is organized in subblocks according to function. The subblocks are Sources, Sinks, Discrete, Linear, Nonlinear, Connections, and Extras. If, for example, we double-click on the Sources block icon, the windows block shown in Figure 1.9 appears on the desktop.

The signal sources are displayed in the window. We can select these sources and drag them from the Sources block window into the system window. Dragging a signal source from Sources into the system window results only in copying the source from one window to another.

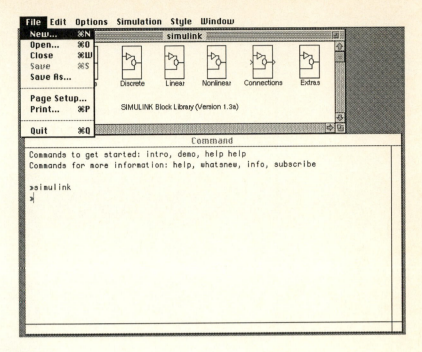

FIGURE 1.7
Select New from the File menu to create the system window.

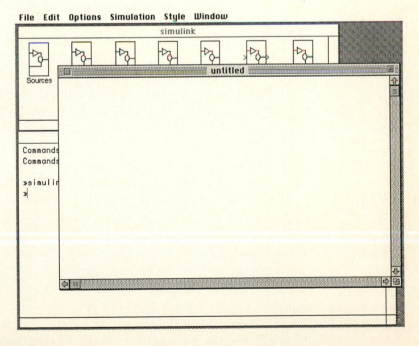

FIGURE 1.8
The new system window created by selecting New from the File menu.

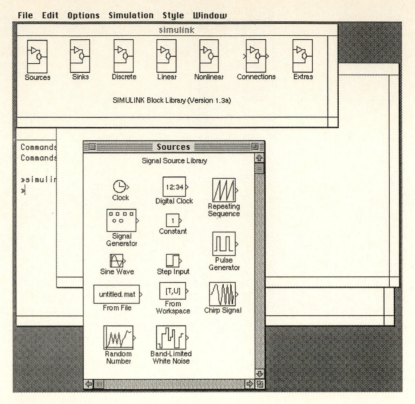

FIGURE 1.9
The Sources block window.

We need to have a way to observe a signal generated by the signal source. The Sinks window (which we open by double-clicking on the Sinks block icon), displays the various instruments for observing the signal, including Scope, Graph, and Auto-Scale Graph, as shown in Figure 1.10. We can use the To Workspace and To File to send write data to a matrix and to a file, respectively, so the matrix data is available in the workspace after the SIMULINK simulation is finished.

After some rearranging of the desktop, the SIMULINK session might resemble Figure 1.11. Notice that the system window has been renamed Test (formerly it was labeled Untitled, as shown in Figure 1.8). We can construct a simple system by dragging the Sine Wave source to the system window and the Auto-Scale Graph instrument from the Sinks window to the test window. The angle bracket ($>$) that appears on the right side of the Sine Wave source is the output port. Similarly, the angle bracket ($>$) that appears on the left side of the Auto-Scale Graph is the input port. To connect the input and output ports, we click on either the input or output port and drag over to the other port. After reaching the other port, release the mouse button to create a connecting line (or data link) between the two ports. When the ports are connected, the angle brackets disappear and a

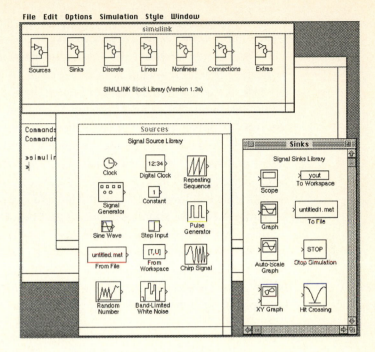

FIGURE 1.10
The Sinks block window.

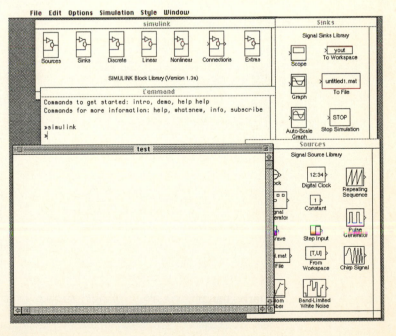

FIGURE 1.11
The system window has been renamed Test.

line with an arrowhead indicating the direction of information flow appears, as shown in Figure 1.12. We are now almost ready to begin a simulation.

We can view and edit the simulation parameters by selecting Parameters from the Simulation menu. A screen similar to the one shown in Figure 1.13 appears. We will definitely want to change the Stop Time since the default value is 999999. Figure 1.13 shows the Stop Time changed to 1000. We can also select the type of numerical integration scheme to use such as Euler, Runge Kutta-23, Runge Kutta-45, and so on. The Runge Kutta-45 is the default integration algorithm. After setting the simulation parameters as we desire, close the dialog box by clicking on OK.

To start the simulation, select Start from the Simulation menu, as shown in Figure 1.14. The simulation begins and the Auto-Scale Graph produces the graph shown in Figure 1.15. The graph is updated dynamically as the data is generated by the Sine Wave source. We can change the parameters of the sine wave (such as frequency) by double-clicking on the Sine Wave source and modifying the desired parameters in the window shown in Figure 1.16. In this case the frequency of the sine wave is 0.05 rad/sec.

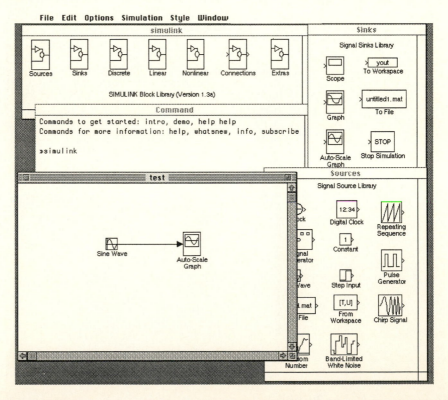

FIGURE 1.12
A simple system with the Sine Wave signal source and the Auto-Scale Graph Sink.

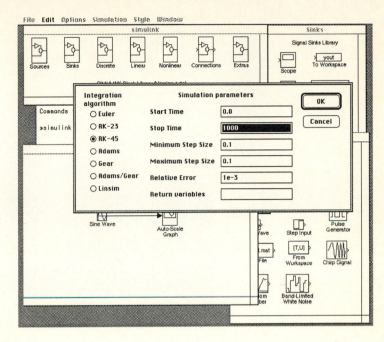

FIGURE 1.13
Initializing the simulation parameters.

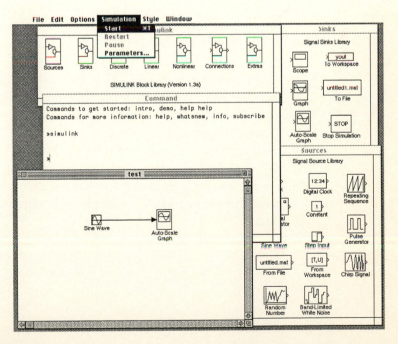

FIGURE 1.14
Starting a simulation by choosing Start under the Simulation menu.

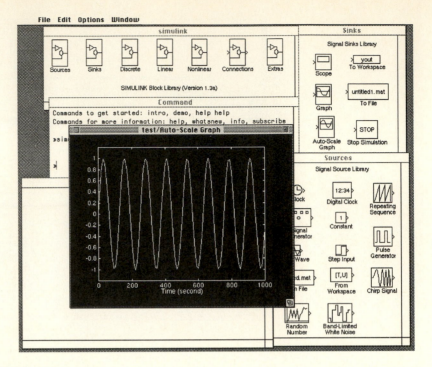

FIGURE 1.15
The simulation session with the the simple simulation system consisting of a Sine Wave signal source and an Auto-Scale Graph for recording the signal. The frequency of the sine wave is $\omega = 0.05$ rad/sec.

The plot shown in Figure 1.15 is the final plot after the simulation stop time has been reached. We can also stop the simulation whenever we choose (before the stop time has been reached) by selecting Stop under the Simulation menu, as shown in Figure 1.17.

Remember to choose Save from the File menu before quitting the SIMULINK session so that all changes and updates to the simulation model are saved for future work. You can quit the SIMULINK session by choosing Close under the File menu and return to a MATLAB command prompt to continue with the MATLAB session.

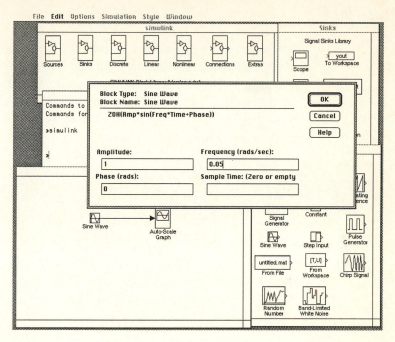

FIGURE 1.16
Setting the sine wave frequency to $\omega = 0.05$ rad/sec.

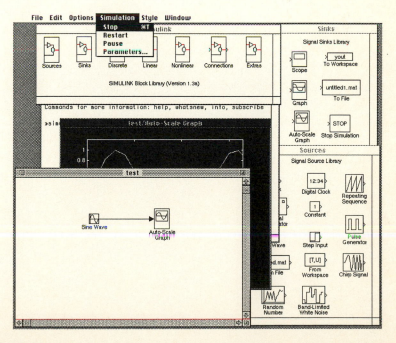

FIGURE 1.17
Stopping a simulation session by choosing Stop under the Simulation menu.

1.4 SUMMARY

In this chapter we provided an introduction to the control design process. A general approach to designing and analyzing a feedback control system was presented and will be followed in subsequent chapters. The space shuttle example was used to motivate the design process and to illustrate the design gap frequently encountered in control system design. The simulation program SIMULINK was also introduced in this chapter.

E X E R C I S E S

E1.1 Develop a SIMULINK simulation in which the signal is a sine wave of frequency 0.1 rad/sec and the output is viewed on an auto-scaled graph.

E1.2 Save the signal values obtained in E1.1 to the MATLAB workspace. Verify that the output has in fact been saved to the workspace by examining the variables exisiting in the MATLAB workspace. (Hint: Use the who function to make the verification.)

E1.3 Use the Signal Generator source to obtain a sine wave of frequency 1 rad/sec, and plot the output on an x-y graph.

CHAPTER 2

Mathematical Models of Systems

Fluid Flow Modeling Example

2.1 Introduction   17

2.2 Fluid Flow Modeling   18

2.3 Mathematical Model and Assumptions   20

2.4 Differential Equations of Motion   22

2.5 Solutions to the Equations of Motion   26

2.6 Summary   33

   Exercises   33

PREVIEW

In this chapter we focus on the modeling of systems. Developing mathematical models is a very important first step in the design process. A fluid flow problem is used to describe a process for obtaining a mathematical model. Obviously the process of modeling is problem dependent, so in this chapter we cannot cover all aspects of the process that may encountered. We do include related topics of linearization, obtaining and using analytic solutions, and computer simulation with MATLAB.

2.1   INTRODUCTION

A critical step in the control system design process is the development of a quantitative mathematical model of the system to be controlled. In many situations the key to effective control system designs is the ability to develop simple models that capture the essential characteristics of the underlying physical systems. Successful control engineers understand physical systems and the associated modeling issues. A working understanding of the governing laws of nature (such as Kirchhoff's voltage law and Newton's laws of translational motion) is important in the modeling process for aerospace, chemical, electrical, industrial, and mechanical engineers in control engineering. It can be a daunting task to make appropriate assumptions in the model development and corresponding linearization assumptions without an understanding of the underlying physical processes.

In this chapter we discuss the development of a linear model representation of a fluid flow system. The approach that we adopt is as follows:

1. Define the system and its components
2. Formulate the mathematical model and list the necessary assumptions
3. Write the differential equations describing the model
4. Identify the input and output variables
5. Solve the equations for the desired output variables
6. Examine the solutions and the assumptions

The components of the design process emphasized in this chapter are shown in Figure 2.1. The central subject of this chapter is obtaining a model of the process. Of course, we will have to establish the system configuration as the starting point; however, this undertaking generally includes a much broader level of activity than we consider now. Since we are concerned here only with obtaining a model of the physical system (in this case a fluid flow system), the issue of sensor and actuator selection and associated modeling is reserved for future chapters. Our primary concern is developing a simple yet accurate model of the physical process and then studying the system through computer simulation using MATLAB.

2.2 FLUID FLOW MODELING

The fluid flow system is shown in Figure 2.2. This problem is adapted from P2.12 in *Modern Control Systems*. The reservoir (or tank) contains water that evacuates through an output port. Water is fed to the reservoir through a pipe controlled by an input valve. The variables of interest are the fluid velocity, V (m/sec), fluid height in the reservoir, H (m), and pressure, p (N/m^2). The pressure is defined as the force per unit area exerted by the fluid on a surface immersed (and at rest with respect to) the fluid. Fluid pressure acts normal to the surface. For further reading on fluid flow modeling, see [9]–[11].

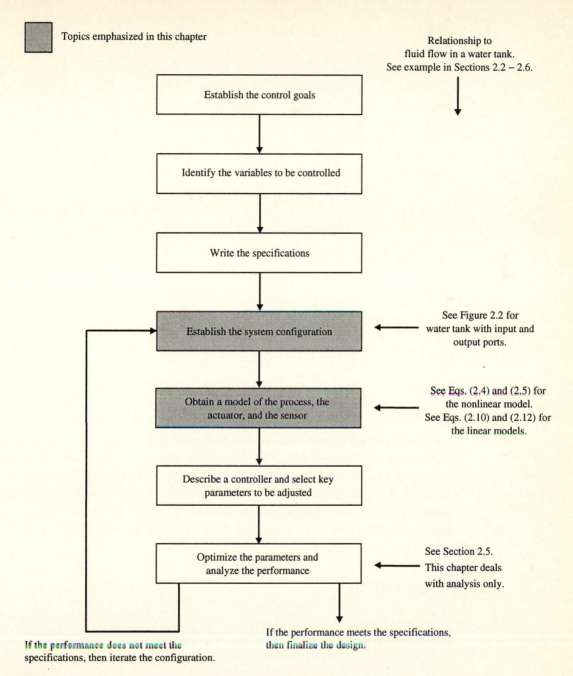

Topics emphasized in this chapter

Relationship to
fluid flow in a water tank.
See example in Sections 2.2 − 2.6.

Establish the control goals

Identify the variables to be controlled

Write the specifications

Establish the system configuration

See Figure 2.2 for
water tank with input and
output ports.

Obtain a model of the process, the
actuator, and the sensor

See Eqs. (2.4) and (2.5) for
the nonlinear model.
See Eqs. (2.10) and (2.12) for
the linear models.

Describe a controller and select key
parameters to be adjusted

Optimize the parameters and
analyze the performance

See Section 2.5.
This chapter deals
with analysis only.

If the performance meets the specifications,
then finalize the design.

If the performance does not meet the
specifications, then iterate the configuration.

FIGURE 2.1
Elements of the control system design process emphasized in this chapter.

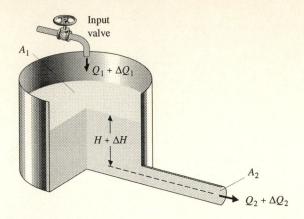

FIGURE 2.2
The fluid flow reservoir configuration.

2.3   MATHEMATICAL MODEL AND ASSUMPTIONS

The general equations of motion (in the context of this chapter) and energy describing fluid flow are quite complicated and not currently used in control system design. The governing equations are coupled nonlinear partial differential equations. We must make some selective assumptions that reduce the complexity of the mathematical model. Although the control engineer is not required to be a fluid dynamicist, and a deep understanding of fluid dynamics is not necessarily acquired during the control system design process, it makes good engineering sense to gain at least a rudimentary understanding of the important simplifying assumptions. For a more complete discussion of fluid motion, see [12]–[14].

2.3.1   Compressibility

We assume that the water in the tank is an incompressible fluid. An **incompressible fluid** has a constant density, ρ (kg/m^3). In fact, all fluids are compressible to some extent. The compressibility factor, k, is a measure of the compressibility of a fluid. A smaller value of k indicates less compressibility. Air (which is a compressible fluid) has a compressibility factor of $k_{air} = 0.98$ m^2/N, while water has a compressibility factor of $k_{H_2O} = 4.9 \times 10^{-10}$ m^2/N $= 50 \times 10^{-6}$ atm^{-1}. In other words, a given volume of water decreases by 50 one-millionths of the original volume for each atmosphere (atm) increase in pressure. Thus the assumption that the water is incompressible is valid for our application.

2.3.2   Viscosity

Consider a fluid in motion. Suppose that initially the flow velocities are different for adjacent layers of fluid. Then an exchange of molecules between the two layers tends to equalize the velocities in the layers. This is internal friction, and the exchange of momentum is known as **viscosity**. Solids are more viscous than fluids, and fluids are more viscous than gases. A measure of viscosity is the coefficient of viscosity, μ (Nsec/m^2). A larger coefficient of viscosity implies higher viscosity. The coefficient of viscosity (under standard conditions, 20° C) for air is

$$\mu_{air} = 0.178 \times 10^{-4} \, Nsec/m^2 \, ,$$

and for water we have

$$\mu_{H_2O} = 1.054 \times 10^{-3} \, Nsec/m^2 \, .$$

So water is about 60 times more viscous than air. Viscosity depends primarily on temperature, not pressure. For comparison, water at 0°C is about 2 times more viscous than water at 20°C. With fluids of low viscosity, such as air and water, the effects of friction are important only in the boundary layer, a thin layer adjacent to the wall of the reservoir and output pipe. We can neglect viscosity in our model development. We say our fluid is **inviscid**.

2.3.3   Irrotational Flow

If each fluid element at each point in the flow has no net angular velocity about that point, the flow is termed **irrotational**. Imagine a paddle wheel immersed in the fluid (say in the output port). If the paddle wheel translates without rotating, the flow is irrotational. We will assume the water in the tank is irrotational. For an inviscid fluid, an initially irrotational flow remains irrotational.

2.3.4   Steady Flow

The water flow in the tank and output port can be either steady or unsteady. The flow is steady if the velocity at each point is constant in time. This does not necessarily imply that the velocity is the same at every point but rather that at any given point the velocity does not change with time. Steady-state conditions can be achieved at low fluid speeds. We will assume **steady flow** conditions. If the output port area is too large, then the flow through the reservoir may not be slow enough to establish the steady-state condition that we are assuming exists and our model will not accurately predict the fluid flow motion.

2.4   DIFFERENTIAL EQUATIONS OF MOTION

The mass of water in the tank at any given time is

$$m = \rho A_1 H , \tag{2.1}$$

where A_1 is the area of the tank, ρ is the water density, and H is the height of the water in the reservoir. The constants for the reservoir system are given in Table 2.1.

In the following formulas, a subscript 1 denotes quantities at the input, and a subscript 2 refers to quantities at the output. Taking the time derivative of m in Eq. (2.1) yields

$$\dot{m} = \rho A_1 \dot{H} ,$$

where we have used the fact that our fluid is incompressible (that is, $\dot{\rho} = 0$) and that the area of the tank, A_1, does not change with time. The change in mass in the reservoir is equal to the mass that enters the tank minus the mass that leaves the tank, or

$$\dot{m} = \rho A_1 \dot{H} = Q_1 - \rho A_2 V_2 , \tag{2.2}$$

where Q_1 is the steady-state input mass flow rate, V_2 is the exit velocity, and A_2 is the output port area. The exit velocity, V_2, is a function of the water height. From Bernoulli's equation [14] we have

$$\frac{1}{2}\rho V_1^2 + P_1 + \rho g H = \frac{1}{2}\rho V_2^2 + P_2 ,$$

where V_1 is the water velocity at the mouth of the reservoir, and P_1 and P_2 are the atmospheric pressures at the input and output, respectively. But P_1 and P_2 are equal to atmospheric pressure, and A_2 is sufficiently small ($A_2 = A_1/100$), so the water flows out slowly and the velocity V_1 is negligible. Thus Bernoulli's equation reduces to

$$V_2 = \sqrt{2gH} . \tag{2.3}$$

TABLE 2.1   Water Tank Physical Constants

ρ (kg/m^3)	g (m/sec^2)	A_1 (m^2)	A_2 (m^2)	H^* (m)	Q^* (kg/sec)
1000	9.8	$\pi/4$	$\pi/400$	1	34.77

Substituting Eq. (2.3) into Eq. (2.2) and solving for $\dot{H}$ yields

$$\dot{H} = -\left[\frac{A_2}{A_1}\sqrt{2g}\right]\sqrt{H} + \frac{1}{\rho A_1}Q_1 . \tag{2.4}$$

Using Eq. (2.3), we obtain the exit mass flow rate

$$Q_2 = \rho A_2 V_2 = \left(\rho\sqrt{2g}A_2\right)\sqrt{H} . \tag{2.5}$$

To keep the equations manageable, we define

$$k_1 \triangleq -\frac{A_2\sqrt{2g}}{A_1} ,$$

$$k_2 \triangleq \frac{1}{\rho A_1} ,$$

$$k_3 \triangleq \rho\sqrt{2g}A_2 .$$

Then we have

$$\dot{H} = k_1\sqrt{H} + k_2 Q_1 , \tag{2.6}$$
$$Q_2 = k_3\sqrt{H} .$$

Eq. (2.6) represents our model of the water tank system, where the input is Q_1 and the output is Q_2. Eq. (2.6) is a nonlinear, first-order, ordinary differential equation model. The nonlinearity comes from the $H^{1/2}$ term. The model in Eq. (2.6) has the functional form

$$\dot{H} = f(H, Q_1) ,$$
$$Q_2 = h(H, Q_1) ,$$

where

$$f(H, Q_1) = k_1\sqrt{H} + k_2 Q_1 \quad \text{and} \quad h(H, Q_1) = k_3\sqrt{H} .$$

2.4.1   Linearization

When the tank system is in equilibrium, we have $\dot{H} = 0$. We can define Q^* and H^* as the equilibrium input mass flow rate and the water level, respectively. The relationship between Q^* and H^* is given by

$$Q^* = -\frac{k_1}{k_2}\sqrt{H^*} = \rho\sqrt{2g}A_2\sqrt{H^*} . \tag{2.7}$$

This condition occurs when just enough water enters the tank in A_1 to make up for the amount leaving through A_2. We can write the water level and input mass flow rate as

$$H = H^* + \Delta H \, , \tag{2.8}$$
$$Q_1 = Q^* + \Delta Q_1 \, ,$$

where ΔH and ΔQ_1 are small deviations from the equilibrium (steady-state) values. The Taylor series expansion about the equilibrium conditions is given by

$$\dot{H} = f(H, Q_1) = f(H^*, Q^*) + \left.\frac{\partial f}{\partial H}\right|_{\substack{H=H^* \\ Q_1=Q^*}} (H - H^*) \tag{2.9}$$

$$+ \left.\frac{\partial f}{\partial Q_1}\right|_{\substack{H=H^* \\ Q_1=Q^*}} (Q_1 - Q^*) + \cdots \, ,$$

where

$$\left.\frac{\partial f}{\partial H}\right|_{\substack{H=H^* \\ Q_1=Q^*}} = \left.\frac{\partial (k_1 \sqrt{H} + k_2 Q_1)}{\partial H}\right|_{\substack{H=H^* \\ Q_1=Q^*}} = \frac{1}{2} \frac{k_1}{\sqrt{H^*}} \, ,$$

and

$$\left.\frac{\partial f}{\partial Q_1}\right|_{\substack{H=H^* \\ Q_1=Q^*}} = \left.\frac{\partial (k_1 \sqrt{H} + k_2 Q_1)}{\partial Q_1}\right|_{\substack{H=H^* \\ Q_1=Q^*}} = k_2 \, .$$

Using Eq. (2.7), we have

$$\sqrt{H^*} = \frac{Q^*}{\rho \sqrt{2g} A_2} \, ,$$

so that

$$\left.\frac{\partial f}{\partial H}\right|_{\substack{H=H^* \\ Q_1=Q^*}} = -\frac{A_2^2}{A_1} \frac{g\rho}{Q^*} \, .$$

It follows from Eq. (2.8) that

$$\dot{H} = \Delta \dot{H} \, ,$$

since H^* is constant. Also, the term $f(H^*, Q^*)$ is identically zero, by definition of the equilibrium condition. Then neglecting the higher order terms in the Taylor series expansion, we have

$$\Delta \dot{H} = -\left[\frac{A_2^2}{A_1} \frac{g\rho}{Q^*}\right] \Delta H + \frac{1}{\rho A_1} \Delta Q_1 \, . \tag{2.10}$$

Eq. (2.10) is a linear model describing the deviation in water level, ΔH, from

the steady-state due to a deviation from the nominal input mass flow rate, ΔQ_1.
Similarly, for the output variable Q_2 we have

$$Q_2 = Q_2^* + \Delta Q_2 = h(H, Q_1) \tag{2.11}$$

$$\approx h(H^*, Q^*) + \left.\frac{\partial h}{\partial H}\right|_{\substack{H=H^* \\ Q_1=Q^*}} \Delta H + \left.\frac{\partial h}{\partial Q_1}\right|_{\substack{H=H^* \\ Q_1=Q^*}} \Delta Q_1 \,,$$

where ΔQ_2 is a small deviation in the output mass flow rate and

$$\left.\frac{\partial h}{\partial H}\right|_{\substack{H=H^* \\ Q_1=Q^*}} = \frac{g\rho^2 A_2^2}{Q^*} \,,$$

and

$$\left.\frac{\partial h}{\partial Q_1}\right|_{\substack{II=II^* \\ Q_1=Q^*}} = 0 \,.$$

Therefore, the linearized equation for the output variable Q_2 is

$$\Delta Q_2 = \left[\frac{g\rho^2 A_2^2}{Q^*}\right] \Delta H \,. \tag{2.12}$$

2.4.2   Transfer Function Models

Taking the time-derivative of Eq. (2.12) and substituting into Eq. (2.10) yields
the input-output relationship

$$\Delta \dot{Q}_2 + \left[\frac{A_2^2}{A_1}\frac{g\rho}{Q^*}\right] \Delta Q_2 = \frac{A_2^2 g\rho}{A_1 Q^*}\Delta Q_1 \,.$$

If we define

$$\Omega \triangleq \frac{A_2^2}{A_1}\frac{g\rho}{Q^*} \,, \tag{2.13}$$

then we have

$$\Delta \dot{Q}_2 + \Omega \Delta Q_2 = \Omega \Delta Q_1 \,. \tag{2.14}$$

Taking the Laplace transform (with **zero initial conditions**) yields the transfer function

$$\Delta Q_2(s)/\Delta Q_1(s) = \frac{\Omega}{s + \Omega} \,. \tag{2.15}$$

We take the Laplace transform of the differential equation with zero initial conditions to obtain a transfer function. A transfer function is always defined with

zero initial conditions.

Eq. (2.15) describes the relationship between the change in the output mass flow rate, $\Delta Q_2(s)$, due to a change in the input mass flow rate, $\Delta Q_1(s)$. We can also obtain a transfer function relationship between the change in the input mass flow rate and the change in the water level in the tank, $\Delta H(s)$. Taking the Laplace transform (with zero initial conditions) of Eq. (2.10) yields

$$\Delta H(s)/\Delta Q_1(s) = \frac{k_2}{s + \Omega} . \tag{2.16}$$

2.5 SOLUTIONS TO THE EQUATIONS OF MOTION

In this section we consider both analytic and numerical solutions to the equations of motion describing the fluid flow in the tank. For the analytic analysis we will consider only the linearized model. Generally we cannot obtain analytic solutions of nonlinear equations of motion. However, using numerical integration methods, we can obtain solutions to both the nonlinear and linear models. In this way we can verify our analytic solutions and compare the linear solutions to the solutions obtained from the nonlinear model. It is then possible to perform computer experiments to investigate the validity of the linearization assumptions.

2.5.1 Analytic Analysis

Given the linear time-invariant model of the water tank system in Eq. (2.14), we can obtain solutions for step and sinusoidal inputs. Remember that our input, $\Delta Q_1(s)$, is actually a change in the input mass flow rate from the steady-state value, Q^*.

Step input Let

$$\Delta Q_1(s) = q_o/s ,$$

where q_o is the magnitude of the step input, and the initial condition is $\Delta Q_2(0) = 0$. Then we can use the transfer function form given in Eq. (2.15) to obtain

$$\Delta Q_2(s) = \frac{q_o \Omega}{s(s + \Omega)} .$$

The partial fraction expansion yields

$$\Delta Q_2(s) = \frac{-q_o}{s + \Omega} + \frac{q_o}{s} .$$

Taking the inverse Laplace transform yields

$$\Delta Q_2(t) = -q_o e^{-\Omega t} + q_o \ .$$

Note that $\Omega > 0$ [see Eq. (2.13)], so the term $e^{-\Omega t}$ approaches zero as t approaches ∞. Therefore, the steady-state output due to the step input of magnitude q_o is

$$\Delta Q_{2_{ss}} = q_o \ .$$

We see that in the steady-state, the deviation of the output mass flow rate from the equilibrium value is equal to the deviation of the input mass flow rate from the equilibrium value. By examining the variable Ω in Eq. (2.13), we find that the larger the output port opening, A_2, the faster the system reaches steady-state. In other words, as Ω gets larger, the exponential term $e^{-\Omega t}$ vanishes more quickly, and steady-state is reached faster.

　　Similarly for the water level we have

$$\Delta H(s) = \frac{-q_o k_2}{\Omega} \left(\frac{1}{s + \Omega} - \frac{1}{s} \right) \ .$$

Taking the inverse Laplace transform yields

$$\Delta H(t) = \frac{-q_o k_2}{\Omega} \left(e^{-\Omega t} - 1 \right) \ .$$

The steady-state change in water level due to the step input of magnitude q_o is

$$\Delta H_{ss} = \frac{q_o k_2}{\Omega} \ .$$

Sinusoidal input　Let

$$\Delta Q_1(t) = q_o \sin \omega t \ ,$$

which corresponds to

$$\Delta Q_1(s) = \frac{q_o \omega}{s^2 + \omega^2} \ .$$

Suppose the system has zero initial conditions, that is, $\Delta Q_2(0) = 0$. Then from Eq. (2.15) we have

$$\Delta Q_2(s) = \frac{q_o \omega \Omega}{(s + \Omega)(s^2 + \omega^2)} \ .$$

Expanding in a partial fraction expansion and taking the inverse Laplace transform yields

$$\Delta Q_2(t) = q_o \Omega \omega \left(\frac{e^{-\Omega t}}{\Omega^2 + \omega^2} + \frac{\sin(\omega t + \phi)}{\omega (\Omega^2 + \omega^2)^{1/2}} \right) ,$$

where $\phi = \tan^{-1}(\omega / \Omega)$. So, as $t \to \infty$, we have

$$\Delta Q_2(t) \quad \to \quad \frac{q_o \Omega}{\sqrt{\Omega^2 + \omega^2}} \sin(\omega t + \phi) .$$

The maximum change in output flow rate is

$$|\Delta Q_2(t)|_{max} = \frac{q_o \Omega}{\sqrt{\Omega^2 + \omega^2}} . \tag{2.17}$$

2.5.2   Computer Analysis

We finish up our investigations of the water tank system by considering a computer analysis. We rely on MATLAB to compute the solution of Eq. (2.10) when the input is a unit step. The nonlinear model describing the water level flow rate is as follows (using the constants given in Table 2.1):

$$\dot{H} = -0.0443\sqrt{H} + 1.2732 \times 10^{-3} Q_1 , \tag{2.18}$$
$$Q_2 = 34.77\sqrt{H} .$$

With $H(0) = 0.5$ m and $Q_1(t) = 34.77$ kg/sec, we can numerically integrate the nonlinear model given by Eq. (2.18) to obtain the time history of $H(t)$ and $Q_2(t)$. The response of the system is shown in Figure 2.3. The subject of numerical integration with MATLAB is beyond the scope of this book; however for the student already familiar with the notion of numerical integration, we include the MATLAB script in Figure 2.3. As expected from Eq. (2.7), the system steady-state water level is $H^* = 1$ m when $Q^* = 34.77$ kg/m^3.

It takes about 250 seconds to reach steady-state. Suppose that the system is at steady-state and we want to evaluate the response to a step change in the input mass flow rate. Consider

$$\Delta Q_1(t) = 1 \text{ kg/sec} .$$

Then we can use the transfer function model to obtain the unit step response with MATLAB. The step response is shown in Figure 2.4 for both the linear and nonlinear models. Using the linear model, we find that the steady-state change in water level is $\Delta H = 5.75$ cm. Using the nonlinear model, we find that the steady-state change in water level is $\Delta H = 5.835$ cm. So we see a small difference in the results obtained from the linear model and the more accurate nonlinear model.

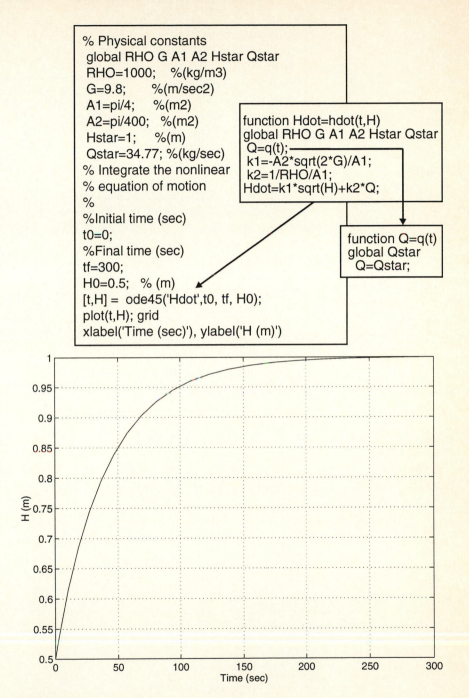

```
% Physical constants
global RHO G A1 A2 Hstar Qstar
RHO=1000;    %(kg/m3)
G=9.8;        %(m/sec2)
A1=pi/4;      %(m2)
A2=pi/400;   %(m2)
Hstar=1;      %(m)
Qstar=34.77; %(kg/sec)
% Integrate the nonlinear
% equation of motion
%
%Initial time (sec)
t0=0;
%Final time (sec)
tf=300;
H0=0.5;   % (m)
[t,H] =  ode45('Hdot',t0, tf, H0);
plot(t,H); grid
xlabel('Time (sec)'), ylabel('H (m)')
```

```
function Hdot=hdot(t,H)
global RHO G A1 A2 Hstar Qstar
Q=q(t);
k1=-A2*sqrt(2*G)/A1;
k2=1/RHO/A1;
Hdot=k1*sqrt(H)+k2*Q;
```

```
function Q=q(t)
global Qstar
Q=Qstar;
```

FIGURE 2.3
The tank water level time history obtained by integrating the nonlinear equations of
motion in Eq. (2.18) with $H(0) = 0.5$ m and $Q_1(t) = Q^* = 34.77$ kg/sec.

```
% Physical constants
global RHO G A1 A2 Hstar Qstar
RHO=1000;   %(kg/m3)
G=9.8;       %(m/sec2)
A1=pi/4; A2=pi/400;   %(m2)
Hstar=1;     %(m)
Qstar=34.7711; %(kg/sec)
% Nonlinear step response
 t0=0;      %Initial time (sec)
 tf=300;    %Final time (sec)
 H0=Hstar;  %Steady-state value (m)
 [tn,H] = ode45('Hdot',t0, tf, H0);
 DHn=H-Hstar;  % Subtract Hstar to obtain DH
% Linear step response
 t=[0:0.1:300];
 k2=1/RHO/A1;
 Omega=A2^2*G*RHO/A1/Qstar;
 num=[k2]; den=[1 Omega];
 DH=step(num,den,t);
% Plot results in centimeters (cm)
plot(t,100*DH,tn,100*DHn,'--'); grid
xlabel('Time (sec)'), ylabel('DH (cm)')
```

```
function Hdot=hdot(t,H)
global RHO G A1 A2 Hstar Qstar
 Q=q(t);
 k1=-A2*sqrt(2*G)/A1;
 k2=1/RHO/A1;
 Hdot=k1*sqrt(H)+k2*Q;
```

```
function Q=q(t)
global Qstar
DQ=1;
Q=Qstar+DQ;
```

See Eq. (2.13)

See Section 2.10 in
Modern Control Systems

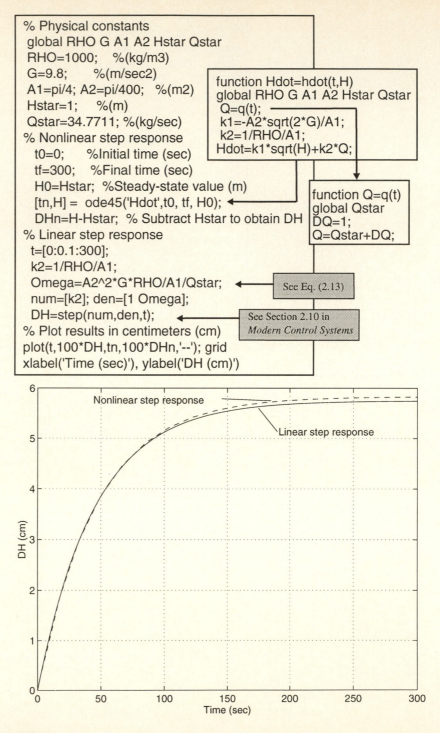

FIGURE 2.4
The step response.

As the final step, we consider the system response to a sinusoidal change in the input flow rate. Let

$$\Delta Q_1(s) = \frac{q_o \omega}{s^2 + \omega^2} \,,$$

where $\omega = 0.05$ rad/sec and $q_o = 1$. The total water input flow rate is

$$Q_1(t) = Q^* + \Delta Q_1(t) \,,$$

where $Q^* = 34.77$ kg/sec. The output flow rate is shown in Figure 2.5.

The response of the water level is shown in Figure 2.6. The water level is sinusoidal, with an average value of $H_{ave} = H^* = 1$ meter. As shown in Eq. (2.17), the output flow rate is sinusoidal in the steady-state, with

$$|\Delta Q_2(t)|_{max} = \frac{q_o \Omega}{\sqrt{\Omega^2 + \omega^2}} = 0.4 \text{ (kg/sec)} \,.$$

Thus in the steady-state (see Figure 2.5) we expect that the output flow rate will oscillate at a frequency of $\omega = 0.05$ rad/sec, with a maximum amplitude of

$$Q_{2_{max}} = Q^* + |\Delta Q_2(t)|_{max} = 35.18 \,.$$

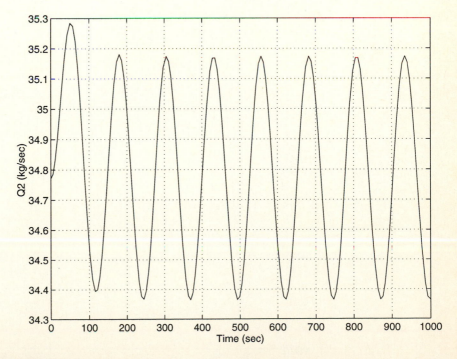

FIGURE 2.5
The output flow rate response to a sinusoidal variation in the input flow.

```
% Physical constants
global RHO G A1 A2 Hstar Qstar
RHO=1000;    %(kg/m3)
G=9.8;        %(m/sec2)
A1=pi/4;      %(m2)
A2=pi/400;    %(m2)
Hstar=1;      %(m)
Qstar=34.77; %(kg/sec)
% Nonlinear sinusoidal response
  t0=0;       %Initial time (sec
  tf=1000;    %Final time (sec)
  H0=Hstar;   %Steady-state value (m)
  [t,H] = ode45('Hdot',t0, tf, H0, 1e-08);
% Compute the output flow rate
k3=RHO*sqrt(2*G)*A2;
Q2=k3*sqrt(H);          ◄── See Eq. (2.5).
% Plot water level in meters (m)
figure(1)
plot(t,H); grid
xlabel('Time (sec)'), ylabel('H (m)')
% Plot output flow rate in kg/sec
figure(2)                ◄── See Figure 2.5.
plot(t,Q2); grid
xlabel('Time (sec)'), ylabel('Q2 (kg/sec)')
```

```
function Q=q(t)
global Qstar
k=1.0;
w=0.05; %Sinusoidal frequency (rad/sec)
%Sinusoidal change in input flow (kg/sec)
DQ=k*sin(w*t);
Q=Qstar+DQ;
```

```
function Hdot=hdot(t,H)
global RHO G A1 A2 Hstar Qstar
Q=q(t);
k1=-A2*sqrt(2*G)/A1;
k2=1/RHO/A1;
Hdot=k1*sqrt(H)+k2*Q;
```

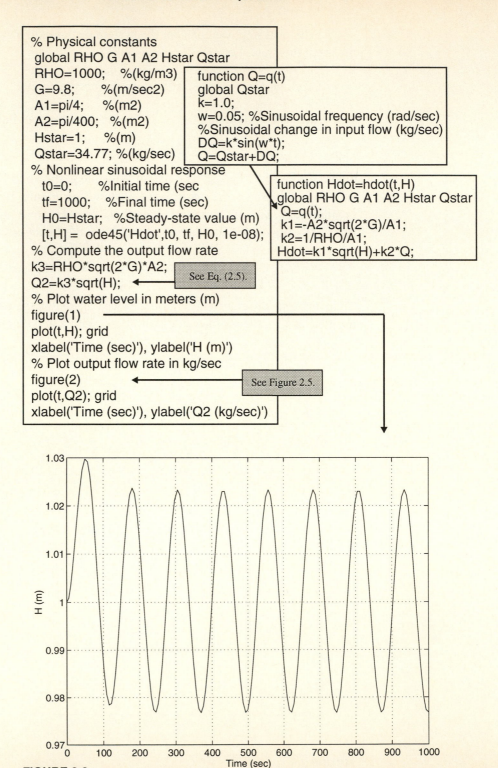

FIGURE 2.6
The water level response to a sinusoidal variation in the input flow.

2.6   SUMMARY

In this chapter we considered some of the issues surrounding the modeling of physical systems. In particular we studied the fluid flow modeling problem. The important assumptions (compressibility, viscosity, irrotationality, and steadiness) were discussed in the context of water flow in a reservoir. The nonlinear equation of motion was developed and subsequently linearized to obtain two linear time-invariant models: one transfer function model for the input mass flow rate, Q_1, to the output, Q_2, and one transfer function model for the input mass flow rate, Q_1, to the water height, H. The linear equations were solved explicitly, and the final steady-state values of the water height and output flow rate were computed for step and sinusoidal inputs. MATLAB was used to obtain step responses for the nonlinear and linear models showing good comparison.

E X E R C I S E S

E2.1   What if the input flow was suddenly stopped? How long would it take to drain the tank? Use the constant values given in Table 2.1. Discuss the validity of the various assumptions made in the derivation (incompressibility, steady flow, and so on) as the water level drops significantly from its steady-state value. Use both analytic methods and MATLAB simulation to strengthen the discussion.

E2.2   Suppose the change in the input flow rate to the tank is a ramp, that is,

$$\Delta Q_1 = \frac{a}{s^2} .$$

The tank is 1.2 m in height and the initial water level is 1 m. How long until the water overflows the tank? Use the constant values given in Table 2.1. Also, let $a = 0.1$ kg/sec^2. Following the spirit of the chapter, first estimate the time to overflow using analytic methods, and then verify using MATLAB simulation.

E2.3   The flow of the water out of the output port can be regulated by a valve. What is the effect of reducing the output flow area from A_2 to $A_2/2$?

CHAPTER 3

State Variable Models

A Space Station Example

3.1 Introduction   35

3.2 Two Simple Physical Systems   35

3.3 Spacecraft Control   43

3.4 Simplified Nonlinear Model   53

3.5 Linearization   54

3.6 Pitch-axis Analysis   56

3.7 Summary   60

Exercises   60

PREVIEW

In this chapter we discuss the mathematical modeling of two simple systems, one mechanical and one electrical. We use MATLAB to analyze the state-space models and obtain step responses for each system. We also look at a more realistic problem: developing a mathematical model of the international space station. Modeling issues we discuss include important assumptions and relevant coordinate frames. We present and linearize the nonlinear model describing the spacecraft attitude motion and momentum storage in the actuators. We also analyze the state-space model for a spacecraft in low-earth orbit. The spacecraft model reflects reality yet retains enough simplicity to be amenable to computer simulation with MATLAB. We show that the linearized model decouples the system into separate channels: one for the pitch axis and one for the coupled roll and yaw axes. We use MATLAB to simulate the pitch axis equations, and we show that, without active control, the space station is unstable in the pitch axis.

3.1   INTRODUCTION

We start the chapter by stating that we cannot adequately address the issue of modeling systems in state variable form in this book. We assume that our years of preparation in the fundamentals—mathematics, physics, chemistry, and so on—have given us the ability to develop mathematical models of systems in our field of interest. In many cases we can develop system models from data derived in the laboratory (or out in the field). We require a mathematical model given in the form of a transfer function or state-space model, and we know that both types of models are equivalent in the sense that they convey the same input and output information. Keep in mind that some branches of control engineering deal with situations for which we cannot obtain accurate mathematical models in any practical way.

A mathematical model is needed for the control system designer to make progress, and yet we cannot spend the time here to cover the issue of modeling very thoroughly—a perplexing problem it seems. Many students become frustrated with textbook problems in which models just appear out of nowhere. Mathematical models for simple systems, such as the rolling cars and the RLC circuit presented in the next section, represent the level of modeling covered in most controls textbooks. Without neglecting the important simple systems (which do appear in real-world problems), we attempt to bring more complicated and relevant modeling problems to the students' attention. The international space station is one such example. The space station model was originally based on state variable methods used in the early Skunk Works days at the NASA Johnson Space Center. We briefly present the model here and discuss many of the interesting associated assumptions and approaches.

The design process is shown in Figure 3.1. The emphasis in this chapter is on obtaining a system model in state variable form.

3.2   TWO SIMPLE PHYSICAL SYSTEMS

In this section we consider one mechanical and one electrical physical system. Most physical systems that we will encounter in practice will be significantly more difficult to model than the ones discussed here; however, the notion of a state vector, inputs, and outputs is generally applicable.

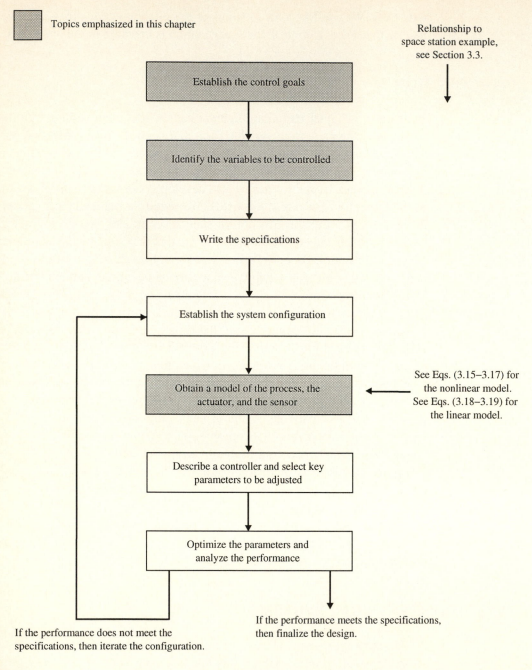

FIGURE 3.1
Elements of the control system design process emphasized in this chapter.

3.2.1 Mechanical: Rolling Carts

Consider the system shown in Figure 3.2. The variables of interest are noted on the figure and defined as

$$M_1, M_2 = \text{mass of cart,}$$
$$p, q = \text{position of cart,}$$
$$u = \text{external force acting on system,}$$
$$k_1, k_2 = \text{spring constant,}$$
$$b_1, b_2 = \text{damping coefficient.}$$

A variety of methods is available for developing the equations of motion for the two cars. Here we use the **free-body diagram** approach. The free-body diagram of mass M_1 is shown in Figure 3.3, where

$$\dot{p}, \dot{q} = \text{velocity of } M_1 \text{ and } M_2, \text{ respectively.}$$

We assume that the cars have negligible rolling friction. We consider any existing rolling friction as lumped into the damping coefficients, b_1 and b_2. When using free-body diagrams we can encounter difficulty when assigning direction to the spring force, $k_1(p - q)$, and damping force, $b_1(\dot{p} - \dot{q})$. Let us consider the mass M_1 first. The position of M_1 is denoted by p. The positive direction

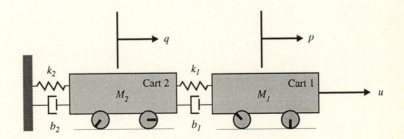

FIGURE 3.2
Two rolling carts attached with springs and dampers.

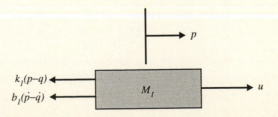

FIGURE 3.3
Free-body diagram of M_1.

of p is specified by the pointing direction of the arrow (see Figure 3.2). Which direction we choose to point the arrow is irrelevant; but once we have specified a direction, the applied forces must be consistent with that direction.

Suppose for example, that we could hold M_2 so that it cannot move (that is, $q = 0$). Then we move the mass M_1 in a positive direction. In this case we selected the positive direction to be to the right. When M_1 is moved to the right, we obtain a reaction force from the spring and damper that resists the motion. The direction of the forces is in the negative direction, or pointing to the left in the free-body diagram. Thus the spring force is

$$f_s = -k_1(p - q),$$

and the damping force is

$$f_d = -b_1(\dot{p} - \dot{q}).$$

Of course, if $q \neq 0$, the spring force is affected. In fact, if $p = q$, then the spring is not compressed or stretched, and the spring force is zero. The above discussion applies to mass M_2 as well. Now, given the free-body diagram with forces and directions appropriately applied, we use Newton's second law (sum of the forces equals mass of the object multiplied by its acceleration) to obtain the equation of motion—one equation for each mass. For mass M_1 we have

$$M_1\ddot{p} = \sum F = u + f_s + f_d = u - k_1(p - q) - b_1(\dot{p} - \dot{q}),$$

or

$$M_1\ddot{p} + b_1\dot{p} + k_1 p = u + k_1 q + b_1\dot{q}, \tag{3.1}$$

where

$$\ddot{p}, \ddot{q} = \text{acceleration of } M_1 \text{ and } M_2, \text{ respectively}.$$

Similarly, for mass M_2 we have

$$M_2\ddot{q} = k_1(p - q) + b_1(\dot{p} - \dot{q}) - k_2 q - b_2\dot{q},$$

or

$$M_2\ddot{q} + (k_1 + k_2)q + (b_1 + b_2)\dot{q} = k_1 p + b_1\dot{p}. \tag{3.2}$$

We now have a model given by the two second-order ordinary differential equations in Eqs. (3.1) and (3.2). We can start developing a state-space model by defining

$$x_1 = p,$$
$$x_2 = q.$$

We could have easily started the process by defining $x_1 = q$ and $x_2 = p$. Clearly the state-space model is not unique.

Denoting the derivatives of x_1 and x_2 as x_3 and x_4, respectively, it follows that

$$x_3 = \dot{x}_1 = \dot{p} \,, \tag{3.3}$$

$$x_4 = \dot{x}_2 = \dot{q} \,. \tag{3.4}$$

Taking the derivative of x_3 and x_4 yields, respectively,

$$\dot{x}_3 = \ddot{p} = -\frac{b_1}{M_1}\dot{p} - \frac{k_1}{M_1}p + \frac{1}{M_1}u + \frac{k_1}{M_1}q + \frac{b_1}{M_1}\dot{q} \,, \tag{3.5}$$

$$\dot{x}_4 = \ddot{q} = \frac{-(k_1 + k_2)}{M_2}q - \frac{(b_1 + b_2)}{M_2}\dot{q} + \frac{k_1}{M_2}p + \frac{b_1}{M_2}\dot{p} \,, \tag{3.6}$$

where we use the relationship for $\ddot{p}$ given in Eq. (3.1) and the relationship for $\ddot{q}$ given in Eq. (3.2). But, $\dot{p} = x_3$ and $\dot{q} = x_4$, so Eq. (3.5) can be written as

$$\dot{x}_3 = -\frac{k_1}{M_1}x_1 + \frac{k_1}{M_1}x_2 - \frac{b_1}{M_1}x_3 + \frac{b_1}{M_1}x_4 + \frac{1}{M_1}u \tag{3.7}$$

and Eq. (3.6) as

$$\dot{x}_4 = \frac{k_1}{M_2}x_1 - \frac{(k_1 + k_2)}{M_2}x_2 + \frac{b_1}{M_2}x_3 - \frac{(b_1 + b_2)}{M_2}x_4 \,. \tag{3.8}$$

In matrix form, Eqs. (3.3), (3.4), (3.7), and (3.8) can be written as

$$\dot{\mathbf{x}} = \mathbf{A}\mathbf{x} + \mathbf{B}u$$

where

$$\mathbf{x} = \begin{pmatrix} x_1 \\ x_2 \\ x_3 \\ x_4 \end{pmatrix} = \begin{pmatrix} p \\ q \\ \dot{p} \\ \dot{q} \end{pmatrix} \,,$$

$$\mathbf{A} = \begin{bmatrix} 0 & 0 & 1 & 0 \\ 0 & 0 & 0 & 1 \\ -\frac{k_1}{M_1} & \frac{k_1}{M_1} & -\frac{b_1}{M_1} & \frac{b_1}{M_1} \\ \frac{k_1}{M_2} & -\frac{(k_1+k_2)}{M_2} & \frac{b_1}{M_2} & -\frac{(b_1+b_2)}{M_2} \end{bmatrix} \,, \quad \text{and} \quad \mathbf{B} = \begin{bmatrix} 0 \\ 0 \\ \frac{1}{M_1} \\ 0 \end{bmatrix} \,,$$

and u is the external force acting on the system (see Figure 3.2). If we choose p as our output, then

$$y = (\ 1 \quad 0 \quad 0 \quad 0\ )\mathbf{x} = \mathbf{Cx}\,.$$

The MATLAB script used to simulate the step response of the rolling car system is shown in Figure 3.4. The parameter values used in the simulation are

$$M_1 = 1\ (\text{slug})\,,$$
$$M_2 = 1\ (\text{slug})\,,$$
$$k_1 = 10\ (\text{lb/ft})\,,$$
$$k_2 = 50\ (\text{lb/ft})\,,$$
$$b_1 = 1\ (\text{lb-sec/ft})\,,$$
$$b_2 = 2\ (\text{lb-sec/ft})\,.$$

When the input is a unit step—that is, $u = 1$ lb—the steady-state position of cart 1 is approximately 1.4 inch. To view the position time-history of cart 2, we can change the output matrix to:

$$\mathbf{C} = (\ 0 \quad 1 \quad 0 \quad 0\ )\,.$$

3.2.2 Electrical: RLC Series Circuit

Consider the RLC series circuit shown in Figure 3.5, where

$$v_{in} = \text{input voltage (volts)}\,,$$
$$L = \text{inductance (H)}\,,$$
$$R = \text{resistance }(\Omega)\,,$$
$$C = \text{capacitance (F)}\,,$$
$$v_c = \text{voltage across the capacitor (volts)}\,,$$
$$i = \text{current (amps)}\,.$$

Using Kirchhoff's voltage law, we obtain

$$L\frac{di}{dt} + Ri + v_c = v_{in}\,, \tag{3.9}$$

where

$$v_c = \frac{1}{C}\int i\,dt\,. \tag{3.10}$$

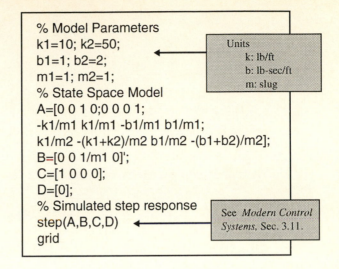

```
% Model Parameters
k1=10; k2=50;
b1=1; b2=2;
m1=1; m2=1;
% State Space Model
A=[0 0 1 0;0 0 0 1;
-k1/m1 k1/m1 -b1/m1 b1/m1;
k1/m2 -(k1+k2)/m2 b1/m2 -(b1+b2)/m2];
B=[0 0 1/m1 0]';
C=[1 0 0 0];
D=[0];
% Simulated step response
step(A,B,C,D)
grid
```

Units
  k: lb/ft
  b: lb-sec/ft
  m: slug

See *Modern Control Systems*, Sec. 3.11.

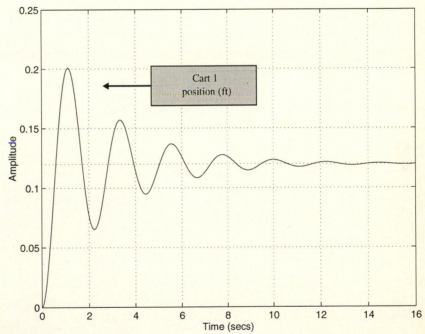

Cart 1 position (ft)

FIGURE 3.4
Rolling cars step response.

We define the state variables as

$$x_1 = v_c \, ,$$
$$x_2 = i \, .$$

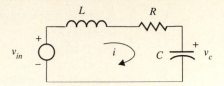

FIGURE 3.5
Simple RLC circuit.

Then taking the derivative of x_1 and using Eq. (3.10) yields

$$\dot{x}_1 = \frac{1}{C}x_2 \, .$$

(3.11)

Also, taking the time derivative of x_2 and using Eq. (3.9) yields

$$\dot{x}_2 = -\frac{R}{L}x_2 - \frac{1}{L}x_1 + \frac{1}{L}v_{in} \, .$$

(3.12)

We can write Eqs. (3.11) and (3.12) in matrix form as

$$\dot{\mathbf{x}} = \mathbf{A}\mathbf{x} + \mathbf{B}u$$

where

$$\mathbf{x} = \begin{pmatrix} x_1 \\ x_2 \end{pmatrix} = \begin{pmatrix} v_c \\ i \end{pmatrix}, \quad u = v_{in} \, ,$$

and

$$\mathbf{A} = \begin{bmatrix} 0 & \frac{1}{C} \\ -\frac{1}{L} & -\frac{R}{L} \end{bmatrix}, \quad B = \begin{bmatrix} 0 \\ \frac{1}{L} \end{bmatrix} \, .$$

With $R = 10 \, \Omega$, $L = 0.2$ H and $C = 0.0015$ F, we have

$$\dot{\mathbf{x}} = \begin{bmatrix} 0 & 666.6 \\ -5 & -50 \end{bmatrix} \mathbf{x} + \begin{bmatrix} 0 \\ 5 \end{bmatrix} u \, .$$

If we can measure v_c, then we have

$$y = \mathbf{C}\mathbf{x} + \mathbf{D}u \, ,$$

where

$$\mathbf{C} = \begin{bmatrix} 1 & 0 \end{bmatrix}, \quad \text{and} \quad \mathbf{D} = [0] \, .$$

We can also compute the transfer function as

$$\frac{v_c(s)}{v_{in}(s)} = G(s) = \mathbf{C}(sI - \mathbf{A})^{-1}\mathbf{B} + \mathbf{D} .$$

So for this case (where we can measure v_c), we have

$$G(s) = \frac{1}{LC} \ \frac{1}{s^2 + \frac{R}{L}s + \frac{1}{LC}} .$$

On the other hand, if we can measure i instead of v_c, we have

$$y = \mathbf{C}\mathbf{x} + \mathbf{D}u ,$$

where

$$\mathbf{C} = [\ 0 \quad 1 \ ] , \quad \text{and} \quad \mathbf{D} = [0] .$$

In this case the transfer function is

$$G(s) = \frac{1}{L} \ \frac{s}{s^2 + \frac{R}{L}s + \frac{1}{LC}} .$$

The MATLAB script used to simulate the step response of the RLC circuit is shown in Figure 3.6. When the input is a unit step—that is $v_{in}(t) = 1$ volt—the steady-state voltage is $v_c(t) = 1$ volt. To view the time-history of the current i, we can change the output matrix in MATLAB script to:

$$\mathbf{C} = [\ 0 \quad 1 \ ] .$$

3.3   SPACECRAFT CONTROL

The international space station, shown in Figure 3.7, is a good example of a multipurpose spacecraft that can operate in many different configurations. A first step in the control system design process is to develop a mathematical model of the spacecraft motion. In general, this model describes the translation and attitude motion of the spacecraft under the influence of external forces and torques, and controller and actuator forces and torques. We must decide on the coordinate systems and notations used in the derivation of the equations of motion. For the spacecraft modeling problem, the relevant coordinate systems can include an inertial reference frame; a spherical coordinate frame; a local vertical/local horizontal frame (or earth-pointing frame); and a body frame attached to the spacecraft. In engineering practice the selection of appropriate coordinate systems is often a time-consuming and (sometimes) contentious matter. This is when we need to use our teaming skills.

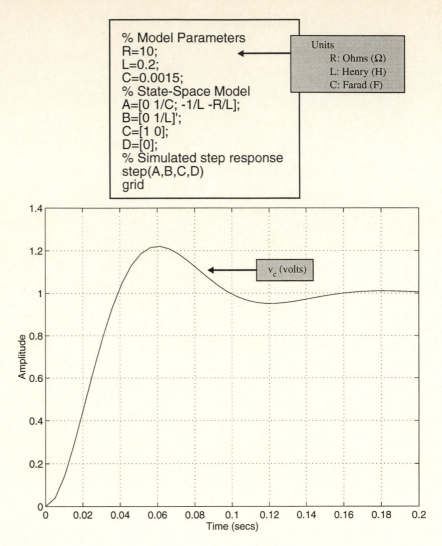

```
% Model Parameters
R=10;
L=0.2;
C=0.0015;
% State-Space Model
A=[0 1/C; -1/L -R/L];
B=[0 1/L]';
C=[1 0];
D=[0];
% Simulated step response
step(A,B,C,D)
grid
```

Units
 R: Ohms (Ω)
 L: Henry (H)
 C: Farad (F)

v_c (volts)

FIGURE 3.6
RLC step response.

Given the appropriate coordinate systems, we can develop the dynamic equations of motion for an orbiting rigid spacecraft. The resulting spacecraft dynamic model is a set of highly coupled, nonlinear ordinary differential equations. Our objective is to simplify the model while retaining important system characteristics. This is not a trivial task, but an important, and often neglected component of control engineering. Unfortunately, in the space we have available, we also must brush over the subject. But in keeping with the theme of this supplement, we recognize the importance of developing simple yet accurate models.

Many spacecraft (such as the international space station) will maintain an earth-pointing attitude. This means that cameras and other scientific instruments

(Courtesy of NASA)

FIGURE 3.7
An artist's conception of the international space station.

pointing down will be able to sense the earth. Conversely scientific instruments pointing up will see deep space, as desired. To achieve earth-pointing attitude, the spacecraft needs an attitude hold control system capable of applying the necessary torques to rotate the vehicle. The international space station will use control moment gyros (CMGs) and reaction control jets as its principal actuators. The CMGs are momentum exchangers and are preferable to reaction control jets because they do not expend fuel. We need to minimize refueling missions to the international space station to the extent possible. However, when the CMGs become saturated, they can no longer provide control torques, and the spacecraft would be uncontrollable without the reaction control jets. We clearly want to avoid CMG saturation. Maintaining the desired attitude requires continuous momentum exchange between the CMGs and the space station because disturbances (such as aerodynamic torques) continuously move the spacecraft away from the desired attitude. Continuous CMG control is not possible without periodic desaturation.

Several methods for desaturating the CMGs are available, but using existing natural environmental torques is the preferred method because it minimizes the use of the reaction control jets. We need a control system that can maintain the desired orientation of the spacecraft, as well as simultaneously perform momentum management for the CMGs. A clever idea is to use gravity gradient torques (which occur naturally and come free of charge) to continuously desaturate the momentum exchange devices. Due to the variation of the earth's gravitational field over the international space station, the total moment generated by the gravitational forces about the spacecraft's center of mass is nonzero. This nonzero moment is called the **gravity gradient torque**. A change in attitude changes the gravity gradient torque acting on the vehicle. Unfortunately, using gravity gradient torques as CMG desaturators can lead to spacecraft orientations away from the desired earth-pointing attitude. Thus combining attitude control and momentum management becomes a problem of compromise.

3.3.1   Inertial Coordinate System

In general, all model development requires the specification of an **inertial** (or nonaccelerating) **reference frame**. The space station is being designed with an inertial frame originating at the center of mass of the earth, as shown in Figures 3.8 and 3.9. The Z_G-axis lies along the polar axis in the direction of the earth's angular momentum vector, where nutation effects of the earth's rotation are neglected. The nutation is a low-amplitude, short-period wobble of the earth's spin axis. The polar axis is assumed to be oriented so that the X_G-axis points from the center of mass of the earth toward the sun. The Y_G-axis is oriented such that the coordinate system is right-handed. The unit vectors, $\mathbf{i}_G, \mathbf{j}_G$, and $\mathbf{k}_G$, are constant, and the origin is nonaccelerating.

3.3.2   Spherical Coordinate System

Consider the spacecraft in orbit about the earth as shown in Figure 3.8. The position of the spacecraft, $\mathbf{r}$, at a given time is given by

$$\mathbf{r} = r\mathbf{i}_S,$$

where r is the radial distance from the center of mass of the earth to the center of mass of the spacecraft, and $\mathbf{i}_S$ is the unit vector pointing in the radial direction. The coordinates given by r, ψ, and ϕ characterize the **spherical coordinate system** originating at the center of mass of the earth. The unit vectors $\mathbf{i}_S, \mathbf{j}_S$, and $\mathbf{k}_S$ are related to the inertially fixed unit vectors $\mathbf{i}_G, \mathbf{j}_G$, and $\mathbf{k}_G$ by the relations

$$\mathbf{i}_S = \cos\psi\cos\phi\,\mathbf{i}_G + \sin\psi\cos\phi\,\mathbf{j}_G + \sin\phi\,\mathbf{k}_G\ ,$$
$$\mathbf{j}_S = -\sin\psi\,\mathbf{i}_G + \cos\psi\,\mathbf{j}_G\ ,$$
$$\mathbf{k}_S = -\cos\psi\sin\phi\,\mathbf{i}_G - \sin\psi\sin\phi\,\mathbf{j}_G + \cos\phi\,\mathbf{k}_G.$$

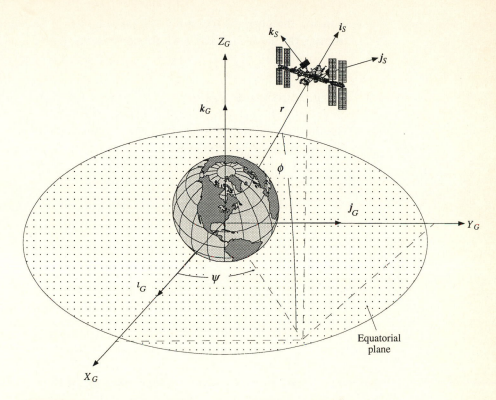

FIGURE 3.8
The inertial and spherical coordinate systems.

We can use these relations to form the transformation matrix

$$
\mathbf{T}^{SG} = \begin{bmatrix}
\cos\psi\cos\phi & \sin\psi\cos\phi & \sin\phi \\
-\sin\psi & \cos\psi & 0 \\
-\cos\psi\sin\phi & -\sin\psi\sin\phi & \cos\phi
\end{bmatrix},
$$

which transforms vectors in the inertial frame to vectors in the spherical frame.

3.3.3 Local Vertical/Local Horizontal Coordinate System

The **local vertical/local horizontal coordinate system** originates at the center of mass of the space station but does not rotate with the body of the spacecraft. As shown in Figure 3.9, the Z_L axis originates at the center of mass of the space station and passes through the center of the earth. The X_L-Z_L plane corresponds to the instantaneous inclined orbit plane. The Y_L axis points in the opposite direction of the instantaneous orbital angular momentum vector. The direction of the X_L axis is chosen to complete the right-handed coordinate system and is in the direction of the space station velocity. When the spacecraft orbit is circular, the X_L axis is aligned with the velocity vector.

3.3.4   Body Coordinate System

The **body coordinate system** originates at the center of mass of the space station and rotates with the spacecraft at angular rate ω. The body frame is related to the local vertical/local horizontal frame by a sequence of three ordered rotations: pitch, yaw, and roll. This (nonunique) sequence of rotations leads to a sequence of transformations from the local vertical/local horizontal frame to the body frame given by

$$\mathbf{T}^{BL} = \mathbf{T}_1\mathbf{T}_3\mathbf{T}_2,$$

where

$$\mathbf{T}_1 = \begin{bmatrix} 1 & 0 & 0 \\ 0 & \cos\theta_1 & \sin\theta_1 \\ 0 & -\sin\theta_1 & \cos\theta_1 \end{bmatrix},$$

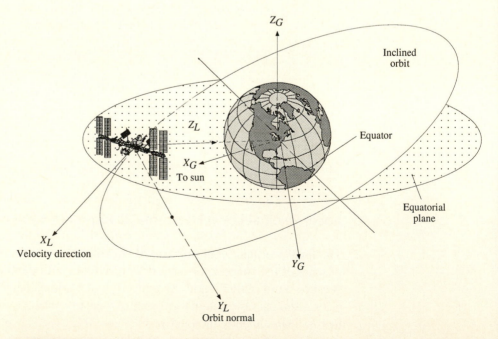

FIGURE 3.9
The inertial and local vertical/local horizontal frames.

$$
\mathbf{T}_3 = \begin{bmatrix} \cos\theta_3 & \sin\theta_3 & 0 \\ -\sin\theta_3 & \cos\theta_3 & 0 \\ 0 & 0 & 1 \end{bmatrix},
$$

$$
\mathbf{T}_2 = \begin{bmatrix} \cos\theta_2 & 0 & -\sin\theta_2 \\ 0 & 1 & 0 \\ \sin\theta_2 & 0 & \cos\theta_2 \end{bmatrix},
$$

and where θ_1 is roll, θ_2 is pitch, and θ_3 is yaw. After multiplying the matrices together, we can write the transformation $\mathbf{T}^{BL}$, which takes vectors in the local vertical/local horizontal frame to vectors in the body frame, as

$$
\mathbf{T}^{BL} = \begin{bmatrix} c\theta_2 c\theta_3 & s\theta_3 & -s\theta_2 c\theta_3 \\ s\theta_1 s\theta_2 - c\theta_1 c\theta_2 s\theta_3 & c\theta_1 c\theta_3 & s\theta_1 c\theta_2 + c\theta_1 s\theta_2 s\theta_3 \\ c\theta_1 s\theta_2 + s\theta_1 c\theta_2 s\theta_3 & -s\theta_1 c\theta_3 & c\theta_1 c\theta_2 - s\theta_1 s\theta_2 s\theta_3 \end{bmatrix}, \qquad (3.13)
$$

where $s\theta \triangleq \sin\theta$ and $c\theta \triangleq \cos\theta$.

When $\theta_1 = \theta_2 = \theta_3 = 0$, the body frame is aligned with the local vertical/local horizontal frame. In this situation the space station is earth-pointing. To maintain an earth-pointing attitude, the international space station must rotate or spin about the Y_B-axis in such a manner that keeps the Z_B-axis pointing towards the earth—in other words, we want to keep the body axes aligned with the local vertical/local horizontal axes.

3.3.5   Some Remarks on External Forces

The spherical frame is useful because we can describe the gravitational forces acting on the international space station very naturally in this reference frame. The force induced by the gravitational field of the earth is the primary force acting on an orbiting spacecraft. Therefore, we cannot ignore gravitational forces in orbital dynamics. However, we can make certain assumptions about the mass distribution of the earth that allow us to simplify the expression for the force due to gravity. We can treat the earth as a point mass. This assumption is equivalent to saying that the earth is a sphere whose mass is distributed uniformly. Actually, /linebreak the earth looks a bit more like a squashed sphere with a bulging equator. For problems requiring highly accurate attitude and trajectory control, we must account for the non-spherical earth with precision. For example, the TOPEX/Poseidon spacecraft radial position was known to within 3–5 cm! Obtaining this precision orbit required a very accurate gravity model. Of course, such accuracy was necessary for that mission because the mission objective was to measure mean sea levels to attempt to characterize global warming. The international space station does not have the same high-accuracy requirements. Using the spherical earth assumption, we can write the gravity force as

$$\mathbf{f}_g = -\frac{\mu m}{r^2}\mathbf{i}_S ,$$

where

$$\mu = 3.986032 \times 10^{14} \, m^3/s^2$$

for the earth and m is the mass of the international space station. The changes in the orbit due to the fact that the earth is not a sphere with uniformly distributed mass are not significant for our attitude control problem. We can assume that a trajectory control system is maintaining the desired orbital trajectory and accounting for nonspherical earth effects.

Over the course of many months, the induced change in the orbit due to the atmospheric drag acting on the international space station is large. Since the attitude control problem considered here deals with time intervals of only a few hours, we can ignore the change in dynamics caused by the atmospheric drag acting on the international space station.

We can also ignore the gravitational effects of the sun and moon and other planets. This is reasonable since we assume that a trajectory control system is handling the orbit control and will account for these disturbances in the orbit. Remember that our main concern is with attitude control and momentum management, not trajectory control. Also, we ignore the effects of the earth's magnetic field, solar radiation and wind, and charged and uncharged particles—they have minor effects on the attitude motion.

Since we want to keep the body axes aligned with the local vertical/local horizontal axes, our control objective is to keep θ_1, θ_2, and θ_3 small, despite the presence of persistent external disturbances. At the same time we must keep the CMGs from saturating. Therefore, the control goal can be stated as

---

Control Goal Minimize the roll, yaw, and pitch angles in the presence of persistent external disturbances while simultaneously minimizing the CMG momentum.

The time rate of change of the angular momentum of a body about its center of mass is equal to the sum of the external torques acting on that body. Thus the attitude dynamics of a spacecraft are driven by externally acting torques. The main external torque acting on the space station is due to gravity. As mentioned previously, the variation in the gravity over the international space station (parts of the vehicle are closer to earth than other parts) generates a moment on the vehicle called a gravity gradient torque.

Since we treat the earth as a point mass, the **gravity gradient torque** [15] acting on the spacecraft is given by

$$\mathbf{T}_g = 3n^2\mathbf{c} \times \mathbf{Ic},\qquad(3.14)$$

where n is the orbital angular velocity ($n = 0.0011$ rad/sec for the space station), and $\mathbf{c}$ is the third column of $\mathbf{T}^{BL}$ given in Eq. (3.13):

$$\mathbf{c} = \begin{bmatrix} -s\theta_2 c\theta_3 \\ s\theta_1 c\theta_2 + c\theta_1 s\theta_2 s\theta_3 \\ c\theta_1 c\theta_2 - s\theta_1 s\theta_2 s\theta_3 \end{bmatrix}.$$

Matrix $\mathbf{I}$ is the spacecraft inertia matrix and is a function of the space station configuration. Thus the gravity gradient torque given in Eq. (3.14) is a function of the spacecraft configuration. For example, suppose that the spacecraft is a sphere with

$$\mathbf{I} = \begin{bmatrix} I_x & 0 & 0 \\ 0 & I_x & 0 \\ 0 & 0 & I_x \end{bmatrix}.$$

Then the gravity gradient torque acting on this spherical spacecraft is zero! Try to show this using Eq. (3.14).

Of course, since we want to use gravity torques to desaturate our CMGs, we want a vehicle configuration capable of generating large gravity gradient torques. Take a look at the international space station shown in Figure 3.9 and decide if it is a good configuration for generating gravity gradient torques.

It also follows from Eq. (3.14) that the gravity gradient torques are a function of the attitude θ_1, θ_2, and θ_3. We want to maintain a prescribed attitude (that is earth-pointing $\theta_1 = \theta_2 = \theta_3 = 0$), but sometimes we must deviate from that attitude so that we can generate gravity gradient torques to assist in the CMG momentum management. Therein lies the conflict; as engineers we often are required to develop control systems to manage conflicts.

Another very important fact is that the gravity gradient torque is not a function of the space station position or velocity. In reality the translational motion is coupled to the attitude motion, and vice versa. All of the simplifying assumptions, however have decoupled the translational motion of the space station from the attitude motion. This decoupling makes things much easier for the attitude control design!

Now we examine the effect of the aerodynamic torque acting on the space station. The aerodynamic torque does affect the attitude of the space station. The aerodynamic torque acting on the space station is generated by the atmospheric drag force that acts through the center of pressure. In general, the center of pressure and the center of mass do not coincide, so aerodynamic torques develop. The aerodynamic torque is a sinusoidal function that tends to oscillate around a small bias. The oscillation in the torque is primarily a result of the earth's diurnal atmospheric bulge. Due to heating, the atmosphere closest to the sun extends further into space than the atmosphere on the side of the earth away from the sun. As the space station travels around the earth (once every 90 minutes or so), it moves through varying air densities, thus causing a cyclic aerodynamic torque. Also, the space station solar panels rotate as they track the sun. This results in another cyclic component of aerodynamic torque. The aerodynamic torque is generally much smaller than the gravity gradient torque. Therefore, for design purposes we can ignore the atmospheric drag torque and view it as a disturbance torque. We would like the controller to minimize the effects of the aerodynamic disturbance on the spacecraft attitude.

Torques caused by the gravitation of other planetary bodies, magnetic fields, solar radiation and wind, and other less significant phenomena are much smaller than the earth's gravity-induced torque and aerodynamic torque. We can ignore these torques in the dynamic model and view them as disturbances.

Finally, we need to discuss the CMGs themselves. First, we will lump all the CMGs together and view them as a single source of torque, instead of considering the motion of each CMG momentum wheel individually. We represent the total CMG momentum with the variable **h**. We need to know and understand the CMG dynamics in the design phase to manage their angular momentum. But since the time constants associated with these dynamics are much shorter than for attitude dynamics, we can ignore CMG dynamics and assume that the CMGs can produce precisely (and without a time-delay) the torque demanded by the control system.

On a practical note, if we insist on including CMG dynamics in the computer simulation of the spacecraft, we would have to use very small time steps in the numerical integration of the equations of motion to see the effects of the CMG dynamics. How small a time step we would need depends on the time constants associated with the CMG dynamics. Experience with a high-fidelity simulation (including nonspherical gravity, aerodynamic torques, and so on hosted in a SIMULINK environment) suggests that a two-hour simulation time can easily turn into an all-day affair. The additional time spent waiting for the simulation to finish is not worth the additional information obtained—at least not in the early design and analysis phase. Ultimately, however, the spacecraft systems must be tested end-to-end by computer simulation. It is not practical or safe to consider testing the international space station control system on-orbit without extensive ground-based computer simulation analysis!

3.4 SIMPLIFIED NONLINEAR MODEL

In the previous sections we discussed numerous assumptions that make the dynamic equations of motion more manageable for design purposes. One result of these assumptions is that the kinematics and dynamics are independent of the orbital (translational) dynamics. Therefore, we can exclude the equations for the translational motion of the spacecraft from the dynamic model for the control and momentum management design problem. A simplified model that we can use as the basis for the control design is

$$\dot{\Theta} = \mathbf{R}\Omega + \mathbf{n}, \tag{3.15}$$

$$\mathbf{I}\dot{\Omega} = -\Omega \times \mathbf{I}\Omega + 3n^2\mathbf{c} \times \mathbf{I}\mathbf{c} - \mathbf{u}, \tag{3.16}$$

$$\dot{\mathbf{h}} = -\Omega \times \mathbf{h} + \mathbf{u}, \tag{3.17}$$

where

$$\mathbf{R}(\Theta) = \frac{1}{c\theta_3} \begin{bmatrix} c\theta_3 & -c\theta_2 s\theta_3 & s\theta_1 s\theta_3 \\ 0 & c\theta_1 & -s\theta_1 \\ 0 & s\theta_1 c\theta_3 & c\theta_1 c\theta_3 \end{bmatrix},$$

$$\mathbf{n} = \begin{bmatrix} 0 \\ n \\ 0 \end{bmatrix}, \quad \Omega = \begin{bmatrix} \omega_1 \\ \omega_2 \\ \omega_3 \end{bmatrix}, \quad \Theta = \begin{bmatrix} \theta_1 \\ \theta_2 \\ \theta_3 \end{bmatrix}, \quad \mathbf{u} = \begin{bmatrix} u_1 \\ u_2 \\ u_3 \end{bmatrix},$$

where u is the CMG input torque. Two good references that describe the fundamentals of spacecraft dynamic modeling are [16] and [17]. Remember that $\mathbf{I}$ represents the inertia matrix, not the identity matrix (see Eq. (3.16)).

Eq. (3.15) represents the kinematics—the relationship between the Euler angles, denoted by Θ, and the body angular velocity vector, Ω. Eq. (3.16) represents the space station attitude dynamics. The terms on the right side represent the sum of the external torques acting on the spacecraft. The first torque is due to inertia cross-coupling, which is a result of the body axes rotating relative to the inertial frame. We prefer to represent the external torques in the spacecraft body frame (since that is the frame in which they naturally occur), so we must account for the relative motion of the body and inertial frames. The second term represents the gravity gradient torque, and the last term is the torque applied to the spacecraft from the actuators, that is the CMGs. The disturbance torques (due to such factors as the atmosphere) are not included in the model presented here but will add directly to the right side of Eq. (3.16). Eq. (3.17) represents the CMG total momentum, in which the time rate of change of the angular momentum contained in the CMGs is equal to $\mathbf{u}$, the interactive torque between the spacecraft and the CMGs. The relative motion of the body and inertial frame is accounted for by the term $-\Omega \times \mathbf{h}$.

How did the mathematical model in Eqs. (3.15)–(3.17) arise? There have been many papers dealing with space station control and momentum management. One of the first to present this nonlinear model is Wie et al. [18]. Other related information about the model and the control problem in general appears in [19]–[23]. Some of these papers may seem intimidating, but take the time to read a few. Articles related to advanced control topics on the space station can be found in [24]–[30]. More recently researchers have started considering developing nonlinear control laws based on the nonlinear model in Eqs. (3.15)–(3.17). Several good articles on this topic appear in [31]–[40].

3.5 LINEARIZATION

A control and momentum management system requires a controller design that can simultaneously prescribe the control $\mathbf{u}$, which maintains the earth-pointing orientation of the spacecraft and minimize $\mathbf{h}$, which is the CMG momentum. The controller should perform both the attitude control and momentum management functions simultaneously. The momentum manager seeks the torque equilibrium attitude (TEA), which makes the sum of the torques acting on the spacecraft zero (or at least minimized). In this attitude the CMG torques are also zero (or minimized). The main factors influencing the TEA are the spacecraft configuration (given by the inertia matrix) and the aerodynamic disturbances. For linearization purposes we assume that the spacecraft has zero products of inertia (that is, the inertia matrix is diagonal) and the aerodynamic disturbances are negligible. Under those conditions, a TEA is

$$\Theta = \mathbf{0}\,,$$
$$\Omega = (0,\ -n,\ 0)\,,$$
$$\mathbf{h} = 0\,.$$

There are other TEA orientations. Try to find them.

The conventional approach to spacecraft momentum management design is to develop a linear model, representing the spacecraft attitude and CMG momentum by linearizing the nonlinear model about the TEA. This linearization is accomplished by a standard Taylor series approximation. Linear control design methods can then be readily applied.

Linearizing the spacecraft nonlinear model in Eqs. (3.15)–(3.17) by assuming small attitude deviations, small rates, small CMG momentum states, zero cross-products of inertia, and negligible aerodynamic disturbances results in a decoupling of the pitch and roll and yaw axes. The linearized equations for the pitch axis are

$$
\begin{bmatrix} \dot{\theta}_2 \\ \dot{\omega}_2 \\ \dot{h}_2 \end{bmatrix} = \begin{bmatrix} 0 & 1 & 0 \\ 3n^2\Delta_2 & 0 & 0 \\ 0 & 0 & 0 \end{bmatrix} \begin{bmatrix} \theta_2 \\ \omega_2 \\ h_2 \end{bmatrix} + \begin{bmatrix} 0 \\ -\frac{1}{I_2} \\ 1 \end{bmatrix} u_2 , \qquad (3.18)
$$

where

$$
\Delta_2 \triangleq \frac{I_3 - I_1}{I_2} .
$$

The subscript 2 refers to the pitch axis terms, the subscript 1 is for the roll axis terms, and 3 is for the yaw axis terms. The linearized equations for the roll/yaw axes are

$$
\begin{bmatrix} \dot{\theta}_1 \\ \dot{\theta}_3 \\ \dot{\omega}_1 \\ \dot{\omega}_3 \\ \dot{h}_1 \\ \dot{h}_3 \end{bmatrix} = \begin{bmatrix} 0 & n & 1 & 0 & 0 & 0 \\ -n & 0 & 0 & 1 & 0 & 0 \\ -3n^2\Delta_1 & 0 & 0 & -n\Delta_1 & 0 & 0 \\ 0 & 0 & -n\Delta_3 & 0 & 0 & 0 \\ 0 & 0 & 0 & 0 & 0 & n \\ 0 & 0 & 0 & 0 & -n & 0 \end{bmatrix} \begin{bmatrix} \theta_1 \\ \theta_3 \\ \omega_1 \\ \omega_3 \\ h_1 \\ h_3 \end{bmatrix}
$$

$$
+ \begin{bmatrix} 0 & 0 \\ 0 & 0 \\ -\frac{1}{I_1} & 0 \\ 0 & -\frac{1}{I_3} \\ 1 & 0 \\ 0 & 1 \end{bmatrix} \begin{bmatrix} u_1 \\ u_3 \end{bmatrix} , \qquad (3.19)
$$

where

$$
\Delta_1 \triangleq \frac{I_2 - I_3}{I_1} \quad \text{and} \quad \Delta_3 \triangleq \frac{I_1 - I_2}{I_3} .
$$

3.6   PITCH-AXIS ANALYSIS

We define the state-vector as

$$\mathbf{x}(t) \triangleq \begin{pmatrix} \theta_2(t) \\ \omega_2(t) \\ h_2(t) \end{pmatrix} ,$$

and the output as

$$y(t) = \theta_2(t) = (\ 1 \quad 0 \quad 0 \ )\mathbf{x}(t) .$$

Here we are considering the spacecraft attitude, $\theta_2(t)$, as the output of interest. We can just as easily look at both the angular velocity, ω_2, and the CMG momentum, h_2, as outputs.

Our state variable model is

$$\dot{\mathbf{x}} = \mathbf{A}\mathbf{x} + \mathbf{B}u , \qquad (3.20)$$
$$y = \mathbf{C}\mathbf{x} + \mathbf{D}u ,$$

where

$$\mathbf{A} = \begin{bmatrix} 0 & 1 & 0 \\ 3n^2\Delta_2 & 0 & 0 \\ 0 & 0 & 0 \end{bmatrix} , \quad \mathbf{B} = \begin{bmatrix} 0 \\ -\frac{1}{I_2} \\ 1 \end{bmatrix} ,$$

$$\mathbf{C} = [\ 1 \quad 0 \quad 0 \ ], \quad \mathbf{D} = [0] ,$$

and where u is the CMG torque in the pitch axis. The solution to the state differential equation, given in Eq. (3.20), is

$$\mathbf{x}(t) = \Phi(t)\mathbf{x}(0) + \int_0^t \Phi(t-\tau)\mathbf{B}u(\tau)d\tau ,$$

where

$$\Phi(t) = \exp(\mathbf{A}t)$$
$$= \mathcal{L}^{-1}\{(s\mathbf{I} - \mathbf{A})^{-1}\}$$

$$
= \begin{bmatrix} \frac{1}{2}(e^{\sqrt{3n^2\Delta_2}t} + e^{-\sqrt{3n^2\Delta_2}t}) & \frac{1}{2\sqrt{3n^2\Delta_2}}(e^{\sqrt{3n^2\Delta_2}t} - e^{-\sqrt{3n^2\Delta_2}t}) & 0 \\ \frac{1}{2}(e^{\sqrt{3n^2\Delta_2}t} - e^{-\sqrt{3n^2\Delta_2}t}) & \frac{1}{2}(e^{\sqrt{3n^2\Delta_2}t} + e^{-\sqrt{3n^2\Delta_2}t}) & 0 \\ 0 & 0 & 1 \end{bmatrix} .
$$

We can see that if $\Delta_2 > 0$, then some elements of the state transition matrix will have terms of the form e^{at}, where $a > 0$. As we shall see, this indicates that our system is unstable. Also, if we are interested in the output, $y(t) = \theta_2(t)$, we have

$$
y(t) = \mathbf{C}\mathbf{x}(t) .
$$

With $\mathbf{x}(t)$ given by

$$
\mathbf{x}(t) = \Phi(t)\mathbf{x}(0) + \int_0^t \Phi(t - \tau)\mathbf{B}u(\tau)d\tau ,
$$

it follows that

$$
y(t) = \mathbf{C}\Phi(t)\mathbf{x}(0) + \int_0^t \mathbf{C}\Phi(t - \tau)\mathbf{B}u(\tau)d\tau .
$$

The transfer function relating the output $Y(s)$ to the input $U(s)$ is

$$
G(s) = \frac{Y(s)}{U(s)} = \mathbf{C}(s\mathbf{I} - \mathbf{A})^{-1}\mathbf{B} = -\frac{1}{I_2(s^2 - 3n^2\Delta_2)} .
$$

The factors of the characteristic equation are

$$
s^2 - 3n^2\Delta_2 = (s + \sqrt{3n^2\Delta_2})(s - \sqrt{3n^2\Delta_2}) = 0 .
$$

If $\Delta_2 > 0$ (that is, if $I_3 > I_1$), then we have two real poles—one in the left half-plane and the other in the right half-plane. The MATLAB script used to simulate the initial condition response of the space station pitch axis when $I_3 > I_1$ as well as the response plot are shown in Figure 3.10. For spacecraft with $I_3 > I_1$, we can say that an earth-pointing attitude is an unstable orientation. This means that active control is necessary.

Conversely, when $\Delta_2 < 0$ (that is, when $I_1 > I_3$), the characteristic equation has two imaginary roots at

$$
s = \pm j\sqrt{3n^2(I_1 - I_2)} .
$$

The MATLAB script used to simulate the initial condition response of the space station pitch axis when $I_1 > I_3$ as well as the response plot are shown in Figure 3.11. This type of spacecraft is marginally stable. In the absence of any CMG torques, the spacecraft will oscillate around the earth-pointing orientation for any small initial deviation from the desired attitude.

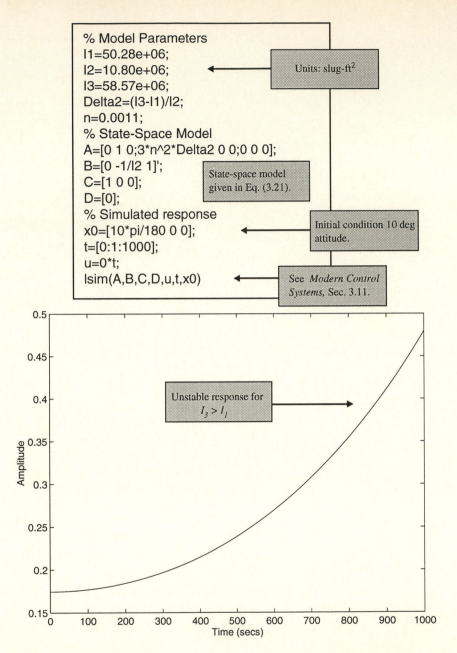

FIGURE 3.10
Initial condition response of the space station pitch axis (without closed-loop control) when $I_3 > I_1$, showing unstable response.

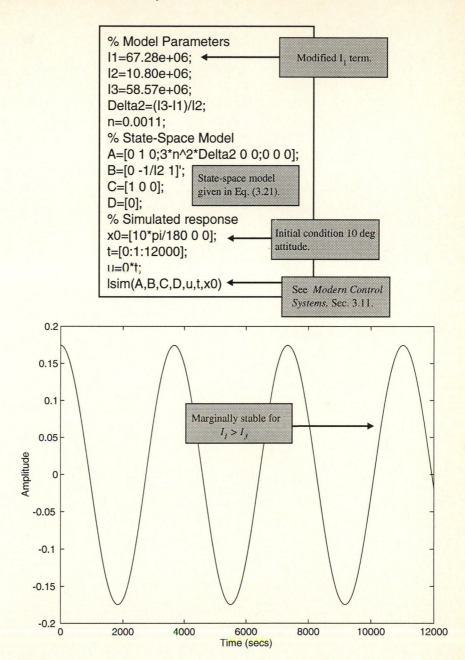

FIGURE 3.11
Initial condition response of the space station when $I_1 > I_3$, showing marginally stable response.

3.7 SUMMARY

In this chapter we discussed mathematical modeling in state variable form of a rolling cart system, an RLC series circuit, and the international space station. We suggested that the issue of system modeling requires a knowledge of fundamentals (mathematics, physics, and so on) in the area of the system of interest. The control system engineer needs a mathematical model to begin the process of designing a controller.

For each system discussed, we used MATLAB to simulate a system response to an input. In the space station example, we showed that the pitch axis motion is unstable when $I_1 < I_3$ and marginally stable when $I_1 > I_3$.

E X E R C I S E S

E3.1 Obtain several of the technical papers cited in Section 3.4, and write a summary of the methodology and findings.

E3.2 Starting from basic principles, derive the rigid body equations of motion given in Eqs. (3.15)–(3.17). State all assumptions clearly and discuss the effect of these assumptions on the model validity.

CHAPTER 4

Feedback Control System Characteristics

Blood Pressure Control Example

4.1 **Introduction** 62

4.2 **Error Signal Analysis** 62

4.3 **Blood Pressure Control During Anesthesia** 71

4.4 **Summary** 83

    Exercises 83

P R E V I E W

In this chapter we use a simple unity feedback system to introduce the sensitivity and complementary sensitivity functions. We discuss the sensitivity of a closed-loop system to plant changes, disturbance rejection, measurement noise attenuation, and steady-state tracking errors. This material reinforces and strengthens the concepts covered in Chapter 4 of *Modern Control Systems*. We investigate the problem of automatic control of blood pressure during anesthesia. We obtain a model for the design and analysis of the feedback control system using experimental impulse response data for a hypothetical patient. The model, representing the patient, describes the relationship between the input anesthesia vapor and the resulting decrease in mean arterial pressure. The impulse response approach illustrated here contrasts with the approaches in Chapters 2 and 3, in which we derived mathematical models from basic physical principles (such as Newton's second law of motion, Kirchhoff's current law, and so on).

4.1   INTRODUCTION

Despite the cost and increased system complexity, closed-loop feedback control has the following advantages:

■ Decreased sensitivity of the system to variations in the parameters of the process $G(s)$

■ Improved rejection of the disturbances

■ Improved measurement noise attenuation

■ Improved reduction of the steady-state error of the system

■ Easy control and adjustment of the transient response of the system

In this chapter we concentrate on the first four items; Chapter 5 deals with the transient response and system performance (the last item on our list) more thoroughly.

At this point we sidestep the central design issues of choosing the controller type and establishing valid values for the controller gains. Instead we provide three proportional-integral-derivative (PID) controllers for blood pressure control. We are interested in the analysis of the feedback control system that uses the candidate PID controllers rather than with the design of the controllers themselves. The analysis of the feedback control system is an important part of the design process. The components of the design process addressed in this chapter are shown in Figure 4.1. The step in the design process involving optimizing the parameters is not emphasized in this chapter since the main concern here is analysis, rather than design. In other words, given three controllers for potential application in the feedback system, we need to decide which one performs better. Of course, defining the term better is sometimes difficult, but we use the term to describe the controller that meets all the design specifications. We provide a PID controller structure and study three reasonable controllers in terms of their respective abilities to reduce system sensitivity, reject disturbances, and reduce steady-state tracking errors.

4.2   ERROR SIGNAL ANALYSIS

The closed-loop feedback control system shown in Figure 4.2 has three inputs—$R(s)$, $D(s)$, and $N(s)$—and one output, $Y(s)$. The signals $D(s)$ and $N(s)$ are the disturbance and measurement noise signals, respectively. The unity feedback system is the most common structure for investigating the attributes of feedback control. Obviously not all feedback control systems have unity feedback (see Section 5.8 in *Modern Control Systems* for a discussion of nonunity feedback systems). However, the conclusions regarding feedback system characteristics that follow from the analysis of unity feedback systems are generally applicable

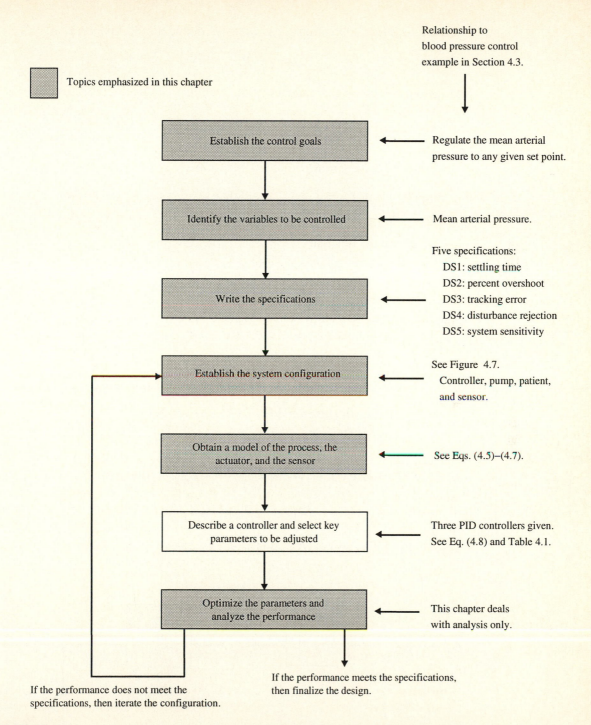

FIGURE 4.1
Elements of the control system design process emphasized in this chapter.

to other feedback configurations as well. We define the tracking error as

$$E(s) = R(s) - Y(s) .$$

After some block diagram manipulation, we find that the output is given by

$$Y(s) = \frac{G_c(s)G(s)}{1 + G_c(s)G(s)} R(s) + \frac{G(s)}{1 + G_c(s)G(s)} D(s) - \frac{G_c(s)G(s)}{1 + G_c(s)G(s)} N(s) .$$

Therefore, with $E(s) = R(s) - Y(s)$, we have

$$E(s) = \frac{1}{1 + G_c(s)G(s)} R(s) - \frac{G(s)}{1 + G_c(s)G(s)} D(s) + \frac{G_c(s)G(s)}{1 + G_c(s)G(s)} N(s) .$$

We define the function

$$L(s) = G_c(s)G(s) ,$$

which is sometimes referred to as the **loop gain**. The function $L(s)$ plays a fundamental role in control system analysis. In terms of $L(s)$ the tracking error is given by

$$E(s) = \frac{1}{1 + L(s)} R(s) - \frac{G(s)}{1 + L(s)} D(s) + \frac{L(s)}{1 + L(s)} N(s) . \qquad (4.1)$$

We can define the function [41]

$$F(s) = 1 + L(s) .$$

Then, in terms of $F(s)$, we define the **sensitivity function** as

$$S(s) = \frac{1}{F(s)} = \frac{1}{1 + L(s)} . \qquad (4.2)$$

This sensitivity function is the same as the one discussed in Section 4.2 of *Mod-*

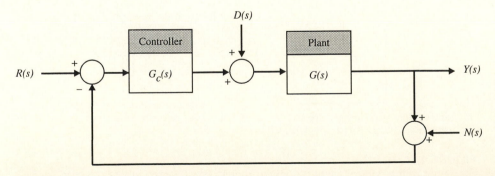

FIGURE 4.2
Unity feedback configuration for error signal analyses.

ern Control Systems. In other words,

$$S(s) = S_G^T(s) \ .$$

The closed-loop transfer function of the system in Figure 4.2 is

$$T(s) = \frac{L(s)}{1 + L(s)} \ .$$

The function $T(s)$ is sometimes called the **complementary sensitivity function**. In terms of the functions $S(s)$ and $T(s)$, we can write the tracking error as

$$E(s) = S(s)R(s) - S(s)G(s)D(s) + T(s)N(s) \ . \tag{4.3}$$

Examining Eq. (4.3), we see that (for a given $G(s)$), if we want to minimize the tracking error, we want both $S(s)$ and $T(s)$ to be small. By small we generally mean small in magnitude (that is, $|T(j\omega)|$ is small over a given range of ω). Remember that $S(s)$ and $T(s)$ are both functions of the controller, $G_c(s)$, which the control design engineer must select. However, the following special relationship between $S(s)$ and $T(s)$ holds

$$S(s) + T(s) = 1 \ .$$

We cannot simultaneously make $S(s)$ and $T(s)$ small. Obviously, design compromises must be made.

4.2.1   Sensitivity Analysis

Suppose the process (or plant) $G(s)$ undergoes a change such that the true plant model is $G(s) + \Delta G(s)$. The change in the plant may be due to a changing external environment or natural aging, or it may just represent the uncertainty in certain plant parameters. We consider the effect on the tracking error $E(s)$ due to $\Delta G(s)$. Relying on the principle of superposition, we can let $D(s) = N(s) = 0$ and consider only the reference input $R(s)$. We will investigate the effects of the disturbance and noise signals on the tracking error in the next sections. From Eq. (4.1), it follows that

$$E(s) + \Delta E(s) = \frac{1}{1 + G_c(s)(G(s) + \Delta G(s))} R(s) \ .$$

Then the change in the tracking error is

$$\Delta E(s) = \frac{-G_c(s)\Delta G(s)}{(1 + G_c(s)G(s) + G_c(s)\Delta G(s))(1 + G_c(s)G(s))} R(s) \ . \tag{4.4}$$

Since we usually find that $G_c(s)G(s) >> G_c(s)\Delta G(s)$, we have

$$\Delta E(s) \approx \frac{-G_c(s)\Delta G(s)}{(1 + L(s))^2} R(s) .$$

We see that the change in the tracking error is reduced by the factor $1 + L(s)$, which is generally greater than 1 over the range of frequencies of interest. We obtained this result in Section 4.2 of *Modern Control Systems*.

For large $L(s)$, we have $1 + L(s) \approx L(s)$, and we can approximate the change in the tracking error by

$$\Delta E(s) \approx -\frac{1}{L(s)} \frac{\Delta G(s)}{G(s)} R(s) .$$

Larger magnitude $L(s)$ translates into smaller changes in the tracking error (that is, **reduced sensitivity** to changes in $\Delta G(s)$ in the process). Also, larger $L(s)$ implies smaller sensitivity, $S(s)$.

From the above discussions, we can state that our goal is to design the controller $G_c(s)$ such that the loop gain is large over the frequencies of interest associated with the expected input reference signals. Since the reference input signals are generally low frequency, we want the loop gain to be large at low frequencies to reduce the sensitivity of the system to changes in the plant. This is equivalent to stating that we want the sensitivity function $S(s)$ small at low frequencies.

Illustrative Example   The control system engineer is generally faced with the situation where the process model $G(s)$ is given and cannot be modified. Design freedom lies in the ability to design and modify the controller $G_c(s)$. This is not the best scenario for the design engineer, however. We would rather see the use of concurrent engineering methods that allow the control system design and plant design to be performed simultaneously. The notion of teaming may eventually be prevalent in engineering, allowing us to address the issues of concurrent plant and control system design.

Consider the case where

$$G_c(s) = K \quad \text{and} \quad G(s) = \frac{ps + 10}{s^2 + 3s + 10} ,$$

and where $p = 1$ is the nominal value. The control engineer must specify the gain K. We want to select a K that can reduce the sensitivity of the closed-loop system to changes in the plant. The frequency plots of the loop gain $L(s)$ and the sensitivity function $S(s)$ are shown in Figures 4.3 and 4.4, respectively, for $K = 1, 5,$ and 10. The loop gain is larger over the range of frequencies for larger values of K, and the sensitivity function $S(s)$ is smaller for the larger values of K. From this we would expect the system with $K = 10$ to be less sensitive to changes in the plant, p. Suppose the plant changes such that

$$\Delta G(s) = \frac{\Delta p s}{s^2 + 3s + 10} = \frac{0.1s}{s^2 + 3s + 10} \, .$$

The change in the tracking error (described in Eq. (4.4)) of a closed-loop system to a sinusoidal input $r(t) = \sin 2t$ is shown in Figure 4.5 for $K = 1, 5,$ and 10. We see that, as expected, the change in the tracking error decreases as K increases.

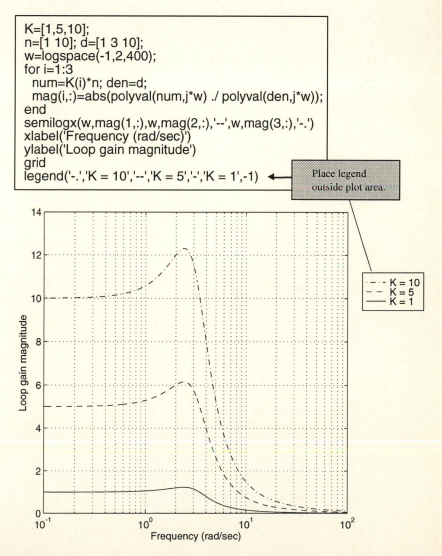

```
K=[1,5,10];
n=[1 10]; d=[1 3 10];
w=logspace(-1,2,400);
for i=1:3
  num=K(i)*n; den=d;
  mag(i,:)=abs(polyval(num,j*w) ./ polyval(den,j*w));
end
semilogx(w,mag(1,:),w,mag(2,:),'--',w,mag(3,:),'-.')
xlabel('Frequency (rad/sec)')
ylabel('Loop gain magnitude')
grid
legend('-.','K = 10','--','K = 5','-','K = 1',-1)
```

Place legend outside plot area.

FIGURE 4.3
Magnitude of $L(s) = K(s + 10)/(s^2 + 3s + 10)$, where $K = 1, 5, 10$.

```
K=[1,5,10];
n=[1 10]; d=[1 3 10];
w=logspace(-1,2,400);
for i=1:3
  num=K(i)*n; den=d;
  [nc,dc]=feedback([1],[1],num,den);
  mag(i,:)=abs(polyval(nc,j*w) ./ polyval(dc,j*w));
end
semilogx(w,mag(1,:),w,mag(2,:),'--',w,mag(3,:),'-.')
xlabel('Frequency (rad/sec)')
ylabel('Sensitivity function magnitude')
grid
legend('-.','K = 10','--','K = 5','-','K = 1',-1)
```

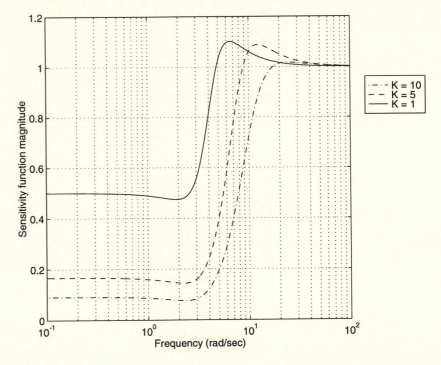

FIGURE 4.4
Sensitivity function $S(s) = 1/(1 + L(s))$, where $L(s) = K(s + 10)/(s^2 + 3s + 10)$ and $K = 1, 5, 10$.

4.2.2   Disturbance Rejection

When $R(s) = N(s) = 0$, it follows from Eq. (4.3) that

$$E(s) = -S(s)G(s)D(s) = -\frac{G(s)}{1 + L(s)}D(s) .$$

```
K=[1,5,10];
dp=0.1; n=[1 3 10 0];
t=[0:0.01:10]; u=sin(2*t);  ◄──── Sinusoidal input.
for i=1:3
  num=-dp*K(i)*n;
  den=[1 6+2.1*K(i) 29+26.3*K(i)+1.1*K(i)^2 60 + ...
      81*K(i)+21*K(i)^2 100+200*K(i)+100*K(i)^2];
  y(:,i)=lsim(num,den,u,t);
end
plot(t,y(:,1),t,y(:,2),'--',t,y(:,3),'-.')
xlabel('Time (sec)')
ylabel('Change in tracking error: Delta e(t)')
grid
legend('-.','K = 10','--','K = 5','-','K = 1',-1)
```

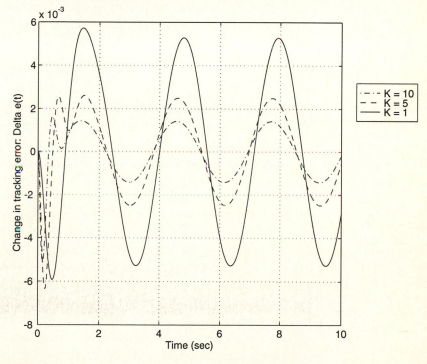

FIGURE 4.5
Response of $\Delta E(s)$, given in Eq. (4.4), to a sinusoidal input $r(t) = \sin 2t$ for $K = 1, 5, 10$ with $\Delta p = 0.1$.

For a fixed $G(s)$ and a given $D(s)$, as the loop gain $L(s)$ increases, the effect of $D(s)$ on the tracking error decreases. In other words, the sensitivity function $S(s)$ is small when the loop gain is large. We say that large loop gain leads to good **disturbance rejection**. More precisely, for good disturbance rejection, we require a large loop gain over the frequencies of interest associated with the expected disturbance signals.

In practice, the disturbance signals are generally low frequency. When that is the case, we say that we want the loop gain to be large at low frequencies. This is equivalent to stating that we want to design the controller $G_c(s)$ such that the sensitivity function $S(s)$ is small at low frequencies.

4.2.3   Measurement Noise Attenuation

When $R(s) = D(s) = 0$, it follows from Eq. (4.3) that

$$E(s) = T(s)N(s) = \frac{L(s)}{1 + L(s)}N(s) \,.$$

As the loop gain $L(s)$ decreases, the effect of $N(s)$ on the tracking error decreases. In other words, the complementary sensitivity function $T(s)$ is small when the loop gain $L(s)$ is small. If we design $G_c(s)$ such that $L(s) << 1$, then the noise is attenuated because

$$T(s) \approx L(s) \,.$$

We say that small loop gain leads to good noise attenuation. More precisely, for effective **measurement noise attenuation**, we need a small loop gain over the frequencies associated with the expected noise signals.

In practice, measurement noise signals are generally high frequency. Thus we want the loop gain to be low at high frequencies. This is equivalent to a small complementary sensitivity function at high frequencies. The separation of disturbances (at low frequencies) and measurement noise (at high frequencies) is very fortunate because it gives the control system designer a way to approach the design process: the controller should be high gain at low frequencies and low gain at high frequencies. Remember that by low and high we mean that the loop gain magnitude is low/high at the various low/high frequencies. It is not always the case that the disturbances are low frequency or that the measurement noise is high frequency. For example, an astronaut pounding on a treadmill on a space station may impart disturbances to the spacecraft at high frequencies. If the frequency separation does not exist, the design process usually becomes more involved (for example, we may have to use notch filters to reject disturbances at known high frequencies).

Considering the previous simple example, we can plot the sensitivity and complementary sensitivity functions, as shown in Figure 4.6 for the case where $K = 10$. If the sensitivity function is small enough at low frequencies to meet the disturbance rejection design specifications (that is, if $K = 10$ is a good value), but the complementary sensitivity function is not small enough at high frequencies to meet the design specifications regarding measurement noise attenuation, then we will have to redesign the controller. We would probably use a more sophisticated controller (such as the PID controller described in the next section).

```
n=10*[1 10];
d=[1 3 10];
[ns,ds]=feedback([1],[1],n,d);
[nt,dt]=cloop(n,d);
w=logspace(-1,2,400);
mags=abs(polyval(ns,j*w) ./ polyval(ds,j*w));
magt=abs(polyval(nt,j*w) ./ polyval(dt,j*w));
semilogx(w,mags,w,magt,'--')
xlabel('Frequency (rad/sec)')
ylabel('S(s) and T(s)')
grid
legend('-','S(s)','--','T(s)',-1)
```

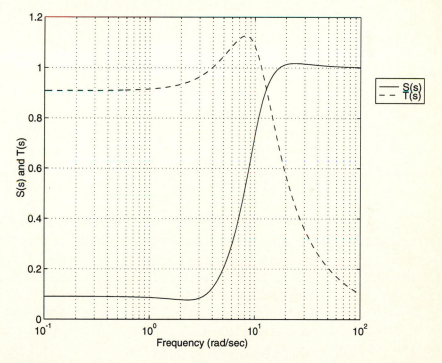

FIGURE 4.6
Sensitivity and complementary sensitivity functions for $L(s) = 10(s + 10)/(s^2 + 3s + 10)$.

4.3   BLOOD PRESSURE CONTROL DURING ANESTHESIA

The objectives of being put to sleep are to eliminate pain, awareness, and natural reflexes so that surgery can be conducted safely. Prior to about 150 years ago, alcohol, opium and cannabis were used to achieve these goals, but they proved inadequate [42]. Pain relief was insufficient both in magnitude and duration: too little pain medication and the patient felt great pain, too much medication and the patient died or became comatose. In the 1850s ether was used successfully in the United States in tooth extractions, and shortly thereafter other means of achieving unconsciousness safely were developed, including the use of chloroform and nitrous oxide.

In a modern operating room, the depth of anesthesia is the responsibility of the anesthetist. Many vital parameters, such as blood pressure, heart rate, temperature, blood oxygenation, and exhaled carbon dioxide, are controlled within acceptable bounds by the anesthetist. Of course, to ensure patient safety, adequate anesthesia must be maintained during the entire surgical procedure. Any assistance that the anesthetist can obtain automatically will increase the safety margins by freeing the anesthetist to attend to other functions not easily automated. This is an example of human computer interaction for the over-all control of a process. Clearly, patient safety is the ultimate objective. Our control goal then is to develop an automated system to regulate the depth of anesthesia. This function is amenable to automatic control and in fact is in routine use in clinical applications [43], [44].

How can we measure the depth of anesthesia? Most anesthetists regard mean arterial pressure (MAP) as the most reliable measure of the depth of anesthesia [45]. The level of the MAP serves as a guide for the delivery of inhaled anesthesia. Based on clinical experience and the procedures followed by the anesthetist, we determine that the variable to be controlled is the mean arterial pressure.

From the control system design perspective, the control goal can be stated in more concrete terms:

Control Goal
Regulate the mean arterial pressure to any desired set-point and maintain the prescribed set-point in the presence of unwanted disturbances.

Associated with the stated control goal, we identify the variable to be controlled:

Variable to Be Controlled
Mean arterial pressure (MAP).

Since it is our desire to develop a system that will be used in clinical applications, it is essential to establish realistic design specifications. In general terms the control system should have **minimal complexity** while satisfying the control specifications. Minimal complexity translates into increased system reliability and decreased cost.

The closed-loop system should respond rapidly and smoothly to changes in the MAP set-point (made by the anesthetist) without excessive overshoot. The closed-loop system should minimize the effects of unwanted disturbances. There are two important categories of disturbances: surgical disturbances, such as skin incisions and measurement errors, such as calibration errors and random stochastic noise. For example, a skin incision can increase the MAP rapidly by 10 mmHg [45]. Finally, since we want to apply the same control system to many different patients and we cannot (for practical reasons) have a separate model for each patient, we must have a closed-loop system that is insensitive to changes in the plant parameters (that is, it meets the specifications for many different people).

Based on clinical experience [43], we can explicitly state the control specifications as follows:

Control Design Specifications

DS1 Settling time less than 20 minutes for a 10% step change from the MAP set-point.

DS2 Percent overshoot less than 15% for a 10% step change from the MAP set-point.

DS3 Zero steady-state tracking error to a step change from the MAP set-point.

DS4 Zero steady-state error to a step surgical disturbance input (of magnitude $|d(t)| \leq 50$) with a maximum response less than $\pm5\%$ of the MAP set-point.

DS5 Minimum sensitivity to plant parameter changes.

We cover the notion of percent overshoot (DS1) and settling time (DS2) more thoroughly in Chapter 5. They fall more naturally in the category of system performance. The remaining three design specifications, DS3–DS5, covering steady-state tracking errors (DS3), disturbance rejection (DS4), and system sensitivity to parameter changes (DS5) are the main topics of this chapter. The last specification, DS5, is somewhat vague, although that is a characteristic of many real-world specifications.

In the system configuration, Figure 4.7, we identify the major system elements as the controller, anesthesia pump/vaporizer, sensor, and patient. This problem is adapted from AP4.5 in *Modern Control Systems*.

The system input, $R(s)$, is the desired mean arterial pressure change, and the output, $Y(s)$, is the actual pressure change. The difference between the desired and the measured blood pressure change forms a signal used by the controller to determine value settings to the pump/vaporizer that delivers anesthesia vapor to the patient.

The model of the pump/vaporizer depends directly on the mechanical design. We will assume a simple pump/vaporizer, where the rate of change of the output vapor is equal to the input valve setting, or

$$\dot{u}(t) = v(t) .$$

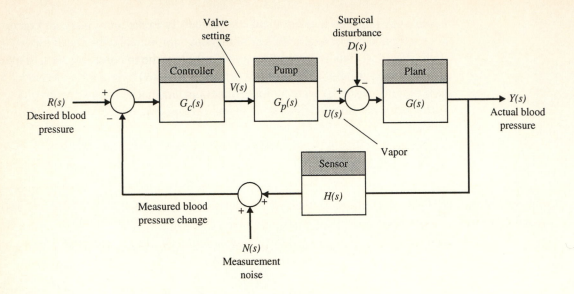

FIGURE 4.7
Blood pressure control system configuration.

The transfer function of the pump is thus given by

$$G_p(s) = \frac{U(s)}{V(s)} = \frac{1}{s}.$$ (4.5)

This is equivalent to saying that, from an input/output perspective, the pump has the impulse response

$$h(t) = 1 \quad t \geq 0.$$

Developing an accurate model of a patient is much more involved. Since the physiological systems in the patient (especially in a sick patient) are not well understood and not easily modeled, a modeling procedure based on knowledge of the underlying physical processes is not practical. Even if such a model could be developed, it would, in general, be a nonlinear, time-varying, multi-input, multi-output model. This type of model is not directly applicable here in our linear, time-invariant, single-input, single-output system setting.

On the other hand, if we view the patient as a system and take an input/output perspective, we can use the familiar concept of an impulse response. Then if we restrict ourselves to small changes in blood pressure from a given set-point (such as 100 mmHg), we might make the case that in a small region around the set-point the patient behaves in a linear time-invariant fashion. This approach fits well into our requirement to maintain the blood pressure around a given set-point (or baseline). The impulse response approach to modeling the patient response to anesthesia has been used successfully in the past [46].

Suppose that we take a black box approach and obtain the impulse response in Figure 4.8 for a hypothetical patient. Notice that the impulse response initially has a time-delay. This reflects the fact that it takes a finite amount of time for the patient MAP to respond to the infusion of anesthesia vapor. We ignore the time-delay in our design and analysis, but we do so with caution. In subsequent chapters we will learn to handle time-delays. So keep in mind that the delay does exist and should be considered in the analysis at some point.

A reasonable fit of the data shown in Figure 4.8 is given by

$$y(t) = te^{-pt} \qquad t \geq 0 ,$$

where $p = 2$ and time (t) is measured in minutes. Different patients are associated with different values of the parameter p. The corresponding transfer function is

$$G(s) = \frac{1}{(s + p)^2} . \tag{4.6}$$

For the sensor we assume a perfect noise-free measurement, so that

$$H(s) = 1 . \tag{4.7}$$

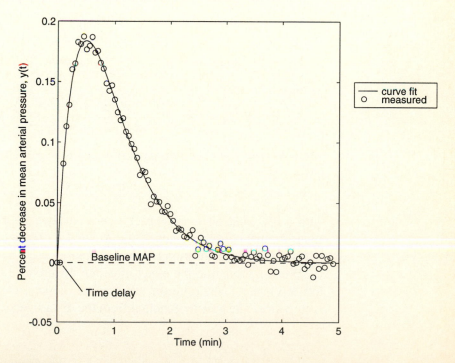

FIGURE 4.8
Mean arterial pressure (MAP) impulse response for a hypothetical patient.

Therefore, we have a unity feedback system.

A good controller for this application is a **proportional-integral-derivative (PID) controller:**

$$G_c(s) = K_1 + sK_2 + \frac{K_3}{s} = \frac{K_2s^2 + K_1s + K_3}{s} , \tag{4.8}$$

where K_1, K_2, and K_3 are the controller gains to be determined to satisfy all design specifications. The selected key parameters are as follows:

Select Key Tuning Parameters   Controller gains K_1, K_2, and K_3.

4.3.1   Steady-state Error Analysis

The tracking error (shown in Figure 4.7 with $D(s) = 0$ and $N(s) = 0$) is

$$E(s) = R(s) - Y(s) = \frac{1}{1 + G_c(s)G_p(s)G(s)}R(s) ,$$

or

$$E(s) = \frac{s^4 + 2ps^3 + p^2s^2}{s^4 + 2ps^3 + (p^2 + K_2)s^2 + K_1s + K_3}R(s) .$$

Using the final-value theorem, we determine that the steady-state tracking error is

$$\lim_{s \to 0} sE(s) = \lim_{s \to 0} \frac{R_0(s^4 + 2ps^3 + p^2s^2)}{s^4 + 2ps^3 + (p^2 + K_2)s^2 + K_1s + K_3} = 0 ,$$

where $R(s) = R_0/s$ is a step input of magnitude R_0. Therefore,

$$\lim_{t \to \infty} e(t) = 0 .$$

With a PID controller, we expect a zero steady-state tracking error (to a step input) for any nonzero values of K_1, K_2, and K_3 since $G_c(s)G_p(s)G(s)$ is type 1. Thus design specification DS3 is satisfied.

When considering the effect of a step disturbance input, we let $R(s) = 0$ and $N(s) = 0$. We desire the steady-state output $Y(s)$ to be zero with a step disturbance. The transfer function from the disturbance $D(s)$ to the output $Y(s)$ is

$$Y(s) = \frac{-G(s)}{1 + G_c(s)G_p(s)G(s)}D(s)$$

$$= \frac{-s^2}{s^4 + 2ps^3 + (p^2 + K_2)s^2 + K_1s + K_3} D(s) \; .$$

When

$$D(s) = \frac{D_0}{s} \; ,$$

we find that

$$\lim_{s \to 0} sY(s) = \lim_{s \to 0} \frac{-D_0 s^2}{s^4 + 2ps^3 + (p^2 + K_2)s^2 + K_1s + K_3} = 0 \; .$$

Therefore,

$$\lim_{t \to \infty} y(t) = 0 \; .$$

Thus a step disturbance of magnitude D_0 will produce no output in the steady-state, as desired.

4.3.2 Sensitivity Analysis

The sensitivity of the closed-loop transfer function to changes in p is given by

$$S_p^T = S_G^T S_p^G \; .$$

We compute S_p^G as follows:

$$S_p^G = \frac{\partial G(s)}{\partial p} \cdot \frac{p}{G(s)} = \frac{-2p}{s + p} \; ,$$

and

$$S_G^T = \frac{1}{1 + G_c(s)G_p(s)G(s)} = \frac{s^2(s + p)^2}{s^4 + 2ps^3 + (p^2 + K_2)s^2 + K_1s + K_3} \; .$$

Therefore,

$$S_p^T = S_G^T S_p^G = -\frac{2p(s + p)s^2}{s^4 + 2ps^3 + (p^2 + K_2)s^2 + K_1s + K_3} \; . \qquad (4.9)$$

We must evaluate the sensitivity function S_p^T, at various values of frequency. For low frequencies we can approximate the system sensitivity S_p^T by

$$S_p^T \approx \frac{2p^2 s^2}{K_3} \; .$$

So at low frequencies and for a given p we can reduce the system sensitivity to variations in p by increasing the PID gain, K_3. Suppose that three PID gain sets have been proposed, as shown in Table 4.1. With $p = 2$ and the PID gains given in Table 4.1, we can plot the magnitude of the sensitivity S_p^T as a function of frequency for each PID controller. The result is shown in Figure 4.9. We see that by using the PID 3 controller with the gains $K_1 = 6$, $K_2 = 4$, and $K_3 = 4$, we have the smallest system sensitivity (at low frequencies) to changes in the plant parameter, p. PID 3 is the controller with the largest gain K_3. As the frequency increases we see in Figure 4.9 that the sensitivity increases, and that PID 3 has the highest peak sensitivity.

4.3.3 Transient Response

Suppose we want to reduce the MAP by a 10% step change. The associated input is

$$R(s) = \frac{R_o}{s} = \frac{10}{s} \, .$$

The step response for each PID controller is shown in Figure 4.10. PID 1 and PID 2 meet the settling time and overshoot specifications; however PID 3 has excessive overshoot. The overshoot is the amount the system output exceeds the desired steady-state response. In this case the desired steady-state response is a 10% decrease in the baseline MAP. When a 15% overshoot is realized, the MAP is decreased by 11.5%, as illustrated in Figure 4.10. The settling time is the time required for the system output to settle within a certain percentage (for example, 2%) of the desired steady-state output amplitude. We cover the notions of overshoot and settling time more thoroughly in Chapter 5. The MATLAB script corresponding to Figure 4.10 is shown in Figure 4.11. The overshoot and settling times are summarized in Table 4.1.

TABLE 4.1   PID Controller Gains and System Performance Results

PID	K_1	K_2	K_3	Input response overshoot (%)	Settling time (min)	Disturbance response overshoot (%)
1	6	4	1	14.0	10.9	5.25
2	5	7	2	14.2	8.7	4.39
3	6	4	4	39.7	11.1	5.16

```
% Patient parameter
  p=2;
% PID Controllers
  K1a=6; K2a=4; K3a=1; %PID 1 gains
  K1b=5; K2b=7; K3b=2; %PID 2 gains
  K1c=6; K2c=4; K3c=4; %PID 3 gains
% Set up vector of s=j*w to evaluate sensitivity
  w=logspace(-1,1,400);
  s=w*j;
% Sensitivity functions with each of the three PID's
  num=2*p*(s+p).*s.^2;
  dena=s.^4+2*p*s.^3+(p^2+K2a)*s.^2+K1a*s+K3a;
  denb=s.^4+2*p*s.^3+(p^2+K2b)*s.^2+K1b*s+K3b;
  denc=s.^4+2*p*s.^3+(p^2+K2c)*s.^2+K1c*s+K3c;
  Sa=abs(num./dena);
  Sb=abs(num./denb);
  Sc=abs(num./denc);
% Generate plot
  semilogx(w,Sa,w,Sb,'--',w,Sc,'-.')
  grid
  xlabel('Frequency (rad/min)')
  ylabel('Sensitivity magnitude')
  legend('-','PID 1','--','PID 2','-.','PID 3',-1)
```

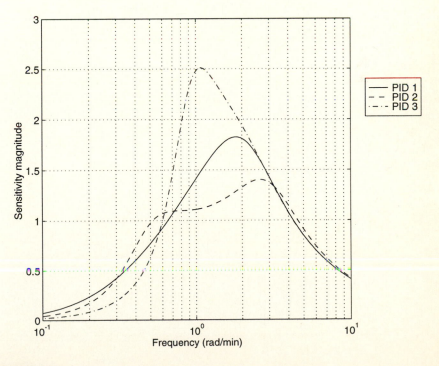

FIGURE 4.9
System sensitivity to variations in the parameter p.

4.3.4   Disturbance Response

From previous analysis we know that the transfer function from the disturbance input $D(s)$ to the output $Y(s)$ is

$$Y(s) = \frac{-G(s)}{1 + G_c(s)G_p(s)G(s)} D(s)$$

$$= \frac{-s^2}{s^4 + 2ps^3 + (p^2 + K_2)s^2 + K_1 s + K_3} D(s) \, .$$

To investigate design specification DS4, we compute the disturbance step response with

$$D(s) = \frac{D_o}{s} = \frac{50}{s} \, .$$

This is the maximum magnitude disturbance ($|d(t)| = D_o = 50$). Since any step disturbance of smaller magnitude (that is, $|d(t)| = D_o < 50$) will result in a smaller maximum output response, we need only to consider the maximum magnitude step disturbance input when determining whether design specification DS4 is satisfied.

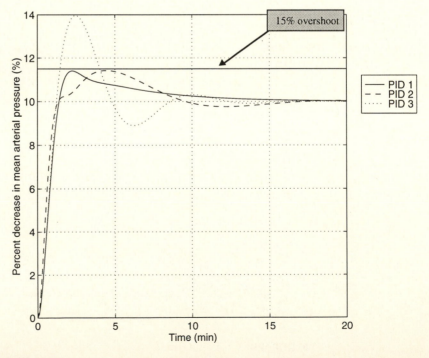

FIGURE 4.10
Mean arterial pressure (MAP) step input response with $R(s) = 10/s$.

```
% Patient Model
p=2;
ng=[1]; dg=[1 2*p p^2];
% PID Controllers + pump model: GcGp(s)
K1a=6; K2a=4; K3a=1; %PID 1 gains
K1b=5; K2b=7; K3b=2; %PID 2 gains
K1c=6; K2c=4; K3c=4; %PID 3 gains
dc=[1 0 0];
nca=[K2a K1a K3a]; ncb=[K2b K1b K3b]; ncc=[K2c K1c K3c];
% Closed-loop transfer functions
[na,d]=series(ng,dg,nca,dc);
[nb,d]=series(ng,dg,ncb,dc);
[nc,d]=series(ng,dg,ncc,dc);
[numa,dena]=cloop(na,d);
[numb,denb]=cloop(nb,d);
[numc,denc]=cloop(nc,d);
% Responses to a step input
tf=20; % maximum time for settling from design spec DS1
t=[0:0.1:tf];
Ro=10;
ya=step(Ro*numa,dena,t);
yb=step(Ro*numb,denb,t);
yc=step(Ro*numc,denc,t);
% Compute percent overshoot and settling times
[POa,TSa]=performance(t,ya,10) % Steady-state value yss=10
[POb,TSb]=performance(t,yb,10)
[POc,TSc]=performance(t,yc,10)  ◄─────────────────┐
% Generate step response plots                     │
plot(t,ya,t,yb,'--',t,yc,':' ,[0 tf],[11.5 11.5]) │
grid                                               │
ylabel('Percent decrease in mean arterial pressure (%)')
legend('-','PID 1','--','PID 2',':','PID 3',-1)
xlabel('Time (min)')    ┌──────────────────────────────────┐
                        │ function [PO,TS] = performance(t,y,yss)
                        │ PO=100*((max(y)-yss)/yss);
                        │ m=find((y-yss)>0.02*yss);
                        │ TS=t(m(length(m)));
                        └──────────────────────────────────┘
```

FIGURE 4.11
MATLAB script used to generate the input response to the step $R(s) = 10/s$ and to compute percent overshoot and settling times.

The unit step disturbance for each PID controller is shown in Figure 4.12. Controller PID 2 meets design specification DS4 with a maximum response less than $\pm 5\%$ of the MAP set-point, while controllers PID 1 and 3 nearly meet the specification. The peak output values for each controller are summarized in Table 4.1.

In summary, given the three PID controllers, we would select PID 2 as the controller of choice. It meets all the design specifications while providing a reasonable insensitivity to changes in the plant parameter.

```
% Patient Model
p=2;
ng=[1]; dg=[1 2*p p^2];
% PID Controllers + pump model: Gc(s)Gp(s)
K1a=6; K2a=4; K3a=1; %PID 1 gains
K1b=5; K2b=7; K3b=2; %PID 2 gains
K1c=6; K2c=4; K3c=4; %PID 3 gains
dc=[1 0 0];
nca=[K2a K1a K3a]; ncb=[K2b K1b K3b]; ncc=[K2c K1c K3c];
% Closed-loop transfer functions from disturbance to output
[numa,dena]=feedback(ng,dg,nca,dc);
[numb,denb]=feedback(ng,dg,ncb,dc);
[numc,denc]=feedback(ng,dg,ncc,dc);
% Step Responses
tf=20; % maximum time for settling
t=[0:0.1:tf];
Do=50; % Do is the disturbance magnitude
ya=step(-Do*numa,dena,t);
yb=step(-Do*numb,denb,t);
yc=step(-Do*numc,denc,t);
% Generate step response plots
plot(t,ya,t,yb,'--',t,yc,':')
grid
ylabel('Percent decrease in mean arterial pressure (%)')
legend('-','PID 1','--','PID 2',':','PID 3',-1)
xlabel('Time (min)')
```

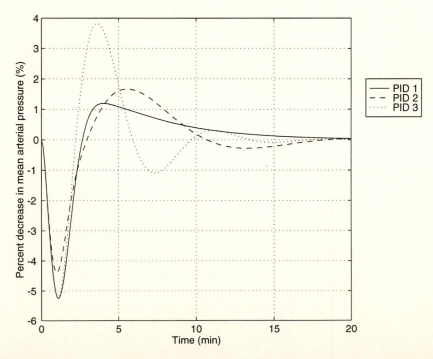

FIGURE 4.12
Mean arterial pressure (MAP) disturbance step response.

4.4 SUMMARY

In this chapter we took a closer look at analysis of the error signal, defined to be the output signal minus the input signal, $E(s) = R(s) - Y(s)$. For the practical situation of low frequency disturbances and high frequency measurement noise, a good design rule-of-thumb is to have a controller that provides high gain at low frequency and low gain at high frequency. Using the problem of blood pressure control during anesthesia as a good example of human-computer interaction to control a process, we analyzed the system response to a step input using three given PID controllers. We found that one PID controller met all the design specifications.

E X E R C I S E S

E4.1 Consider parameter p in the patient transfer function model $G(s)$. For PID 2, how much can p vary and the closed-loop system remain stable? How much can p vary and the closed-loop system meet the settling time and percent overshoot specifications given in Section 4.3?

E4.2 Suppose the pump/vaporizer is more accurately modeled as a constant gain,

$$G_p(s) = K_p ,$$

where $K_p = 0.1$. Design a PID controller to meet the control design specifications given in Section 4.3.

E4.3 Suppose the pump/vaporizer is more accurately modeled as

$$G_p(s) = \frac{1}{s(s+1)} .$$

Design a controller to meet the control design specifications given in Section 4.3.

E4.4 The sensor dynamics are most accurately modeled by

$$H(s) = \frac{2}{s+2} .$$

How does this affect the closed-loop system response with PID 2 in the loop?

CHAPTER 5

Performance of Feedback Control Systems

Airplane Lateral Dynamics Example

5.1 Introduction   85

5.2 Airplane Lateral Dynamics   86

5.3 Bank Angle Control Design   91

5.4 Simulation Development   95

5.5 Summary   102

   Exercises   102

P R E V I E W

In this chapter we begin our investigation of the performance of feedback control systems that will continue in the remaining chapters. To solve the problem of bank angle control of an aircraft we use SIMULINK to build a reasonably simple yet effective model of the response to aileron deflections. Then using the idea of model simplification, we develop a second-order approximate model. The second-order approximation gives us insight into the expected behavior of the system and how to obtain an initial design. Finally we use a SIMULINK simulation to perform analysis in an interactive mode.

5.1 INTRODUCTION

Each time we fly on a commercial airliner, we experience first-hand the benefits of automatic control systems. These systems assist pilots by providing pilot relief (for such emergencies as going to the restroom) during extended flights and by improving the handling qualities of the aircraft over a wide range of flight conditions. The special relationship between flight and controls began in the early work of the Wright brothers. Using wind tunnels the Wright brothers applied systematic design techniques to make their dream of powered flight a reality. This systematic approach to design was partially responsible for their success.

Another significant aspect of their approach was their emphasis on flight controls; the brothers insisted that their aircraft be pilot controlled. Observing birds control their rolling motion by twisting their wings, the Wright brothers built aircraft with mechanical mechanisms that twisted their airplane wings. Today we no longer use wing warping as a mechanism for performing a roll maneuver, instead we control rolling motion by using ailerons, as shown in Figure 5.1. The Wright brothers also used elevators (located forward) for longitudinal control (pitch motion) and rudders for lateral control (yaw motion). Today's aircraft still use both elevators and rudders, although the elevators are generally located on the tail (rearward).

The first controlled, powered, unassisted take-off flight occurred in 1903 with the Wright Flyer I (a.k.a. Kitty Hawk). The first practical airplane, the Flyer III could fly figure eights and stay aloft for half an hour. Three-axis flight control was a major (and often overlooked) contribution of the Wright brothers. A concise historical perspective is presented in Stevens and Lewis [48].

The continuing desire to fly faster, lighter, and longer fostered further developments in automatic flight control. Today's challenge is to develop a single-stage-to-orbit aircraft/spacecraft that can take off and land on a standard runway. Maybe you will have the chance to make this a reality!

The main topic of this chapter is control of the automatic rolling motion of

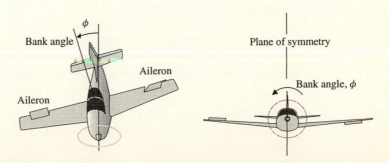

FIGURE 5.1
Control of the bank angle of an airplane using differential deflections of the ailerons.

an airplane. We introduce the use of simulations using SIMULINK to assist us in design and analysis processes, allowing us to perform design iterations easily. The elements of the design process emphasized in this chapter are illustrated in Figure 5.2.

5.2   AIRPLANE LATERAL DYNAMICS

As are mathematical models for many real systems, an accurate mathematical model describing the motion (translational and rotational) of an aircraft is a complicated set of highly nonlinear, time-varying, coupled differential equations. A good description of the process of developing such a mathematical model appears in Etkin and Reid [49].

For our purposes a simplified dynamic model is required for the autopilot design process. A simplified model might consist of a transfer function describing the input/output relationship between the aileron deflection and the aircraft bank angle. Obtaining such a transfer function would require many prudent simplifications to the original high-fidelity, nonlinear mathematical model.

Suppose we have a rigid aircraft with a plane of symmetry (see Figure 5.1). This problem is adapted from DP5.1 in *Modern Control Systems*. The airplane is assumed to be cruising at subsonic or low supersonic (Mach < 3) speeds. This allows us to make a flat-earth approximation. We ignore any rotor gyroscopic effects due to spinning masses on the aircraft (such as propellors or turbines). These assumptions allow us to decouple the longitudinal rotational (pitching) motion from the lateral rotational (rolling and yawing) motion.

Of course, we also need to consider a linearization of the nonlinear equations of motion. To accomplish this, we consider only steady-state flight conditions such as

- Steady, wings-level flight
- Steady, level turning flight
- Steady, symmetric pull-up
- Steady roll

For the problem in this chapter, we assume that the airplane is flying at low speed in a steady, wings-level attitude, and we want to design an autopilot to control the rolling motion. We can state the control goal as follows:

Control Goal   Regulate the airplane bank angle to zero degrees (steady, wings-level) and maintain the wings-level orientation in the presence of unpredictable external disturbances.

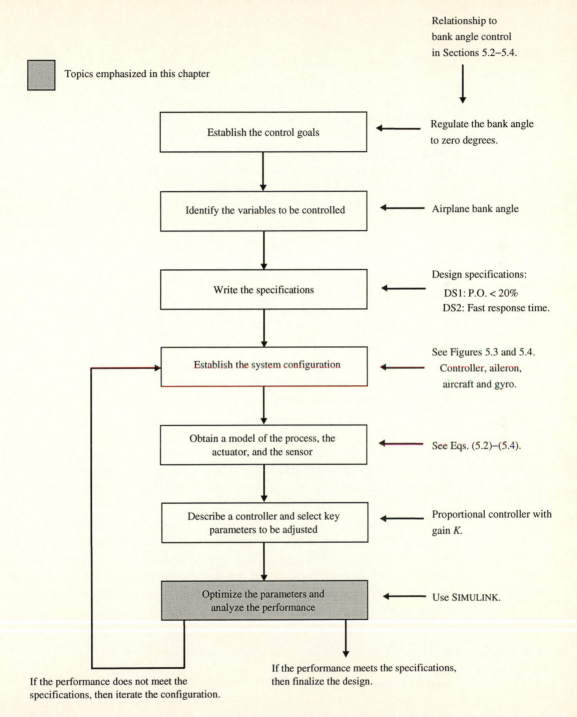

Topics emphasized in this chapter

Relationship to
bank angle control
in Sections 5.2–5.4.

Establish the control goals

Regulate the bank angle
to zero degrees.

Identify the variables to be controlled

Airplane bank angle

Write the specifications

Design specifications:
   DS1: P.O. < 20%
   DS2: Fast response time.

Establish the system configuration

See Figures 5.3 and 5.4.
   Controller, aileron,
   aircraft and gyro.

Obtain a model of the process, the
actuator, and the sensor

See Eqs. (5.2)–(5.4).

Describe a controller and select key
parameters to be adjusted

Proportional controller with
gain K.

Optimize the parameters and
analyze the performance

Use SIMULINK.

If the performance does not meet the
specifications, then iterate the configuration.

If the performance meets the specifications,
then finalize the design.

FIGURE 5.2
Elements of the control system design process emphasized in this chapter.

We identify the variable to be controlled as

Airplane bank angle (denoted by ϕ).

Defining system specifications for aircraft control is complicated, so we do not attempt it here. It is a subject in and of itself, and many engineers have spent significant efforts developing good, practical design specifications. The goal is to design a control system such that the dominant closed-loop system poles have satisfactory natural frequency and damping [48]. How do we define satisfactory? On what test input signals do we base our analysis?

The *Cooper-Harper* pilot opinion ratings provide a way to correlate the feel of the airplane to control design specifications [50]. These ratings address the handling qualities issues. Many flying qualities requirements are specified by government agencies, such as the United States Air Force [51]. The USAF MIL-F-8785C is a source of time-domain control system design specifications.

For example we might design an autopilot control system for an aircraft in steady, wings-level flight to achieve a 20% overshoot to a step input with minimal oscillatory motion and rapid response time (that is, a short time-to-peak). Subsequently we implement the controller in the aircraft control system and conduct flight tests or high-fidelity computer simulations, after which the pilots tell us whether they liked the performance of the aircraft. If the overall performance was not satisfactory, we change the time-domain specification (in this case a percent overshoot specification) and redesign until we achieve a feel and performance that pilots (and ultimately passengers) will accept. Despite the simplicity of this approach and many years of research, precise control system design specifications that provide acceptable airplane flying characteristics in all cases are still not available [48].

The control design specifications given in this chapter may seem somewhat contrived. They are! In reality the specifications would have to be much more involved and, in many ways, much less precise. But recall in Chapter 1 we discussed the fact that we must begin the design process somewhere. With that approach in mind, we select simple design specifications and begin the iterative design process using SIMULINK. The design specifications are

DS1 Percent overshoot less than 20% for a unit step input.

DS2 Fast response time as measured by time-to-peak.

Building a realistic mathematical model of aircraft motion is a complicated process. By making the simplifying assumptions discussed above and linearizing about the steady, wings-level flight condition, we can obtain a transfer function model describing the bank angle output, $\phi(s)$, to the aileron deflection input, $\delta_a(s)$. The transfer function has the form

$$\frac{\phi(s)}{\delta_a(s)} = \frac{k(s - c_o)(s^2 + b_1 s + b_o)}{s(s + d_o)(s + e_o)(s^2 + f_1 s + f_o)} . \tag{5.1}$$

The lateral (roll/yaw) motion has three main modes—Dutch roll mode, spiral mode, and roll subsidence mode. The **Dutch roll** mode, which gets its name from its similarities to the motion of an ice speed skater, is characterized by a rolling and yawing motion. The airplane center of mass follows nearly a straight-line path, and a rudder impulse can excite this mode. The **spiral** mode is characterized by a mainly yawing motion with some roll motion. This is a weak mode, but it can cause an airplane to enter a steep spiral dive. The **roll subsidence** motion is almost a pure roll motion. This is the motion we are concerned with for our autopilot design. The denominator of the transfer function in Eq. (5.1) shows two first-order modes (spiral and roll subsidence modes) and a second-order mode (Dutch roll mode).

In general the coefficients, c_o, b_o, b_1, d_o, e_o, f_o, f_1, and the gain, k, are complicated functions of stability derivatives. The stability derivatives are functions of the flight conditions and the aircraft configuration; they differ for different aircraft types. The coupling between the roll and yaw is included in Eq. (5.1).

In the transfer function in Eq. (5.1), the pole at $s = -d_o$ is associated with the spiral mode. The pole at $s = -e_o$ is associated with the roll subsidence mode. Generally, $e_o >> d_o$. For an F-16 flying at 500 ft/sec in steady, wings-level flight, we have $e_o = 3.57$ and $d_o = 0.0128$ [48]. The complex conjugate poles given by the term $s^2 + f_1 s + f_o$ represent the Dutch roll motion.

For low angles of attack (such as with steady, wings-level flight), the Dutch roll mode generally cancels out of the transfer function. This is an approximation, but it is consistent with our other simplifying assumptions. Also, we can ignore the spiral mode since it is essentially a yaw motion only weakly coupled to the roll motion. The zero at $s = c_o$ represents a gravity effect that causes the aircraft to sideslip as it rolls. We are assuming that this effect is negligible, since it is most pronounced in a slow roll maneuver in which the sideslip is allowed to build up, and we assume that the aircraft sideslip is small or zero. Therefore we can simplify the transfer function in Eq. (5.1) to obtain a single-degree-of-freedom approximation:

$$\frac{\phi(s)}{\delta_a(s)} = \frac{k}{s(s + e_o)} . \tag{5.2}$$

For our aircraft we select $e_o = 1.4$ and $k = 11.4$. The associated time-constant of the roll subsidence is $\tau = 1/e_o = 0.7$ second. These values represent a fairly fast rolling motion response typical of a fighter aircraft.

For the aileron actuator model, we typically use a simple first-order system model,

$$\frac{\delta_a(s)}{e(s)} = \frac{p}{s+p} \, , \tag{5.3}$$

where $e(s) = \phi_d(s) - \phi(s)$. In this case we select $p = 10$. This corresponds to a time-constant of $\tau = 1/p = 0.1$ second. This is a typical value consistent with a fast response. We need to have an actuator with a fast response so that the dynamics of the actively controlled airplane will be the dominant component of the system response. A slow actuator is akin to a time-delay that can cause performance and stability problems.

For a high-fidelity simulation, we would need to develop an accurate model of the gyro dynamics. The gyro, typically an integrating gyro, is usually characterized by a very fast response. To remain consistent with our other simplifying assumptions, we ignore the gyro dynamics in the design process. This means we assume that the sensor measures the bank angle precisely. The gyro model is given by a unity transfer function,

$$K_g = 1 \, . \tag{5.4}$$

Thus our physical system model is given by Eqs. (5.2), (5.3), and (5.4).

The controller we select for this design is a proportional controller,

$$G_c(s) = K \, .$$

The system configuration is shown in Figure 5.3 . The select key parameter is as follows:

Select Key Tuning Parameter   Controller gain K.

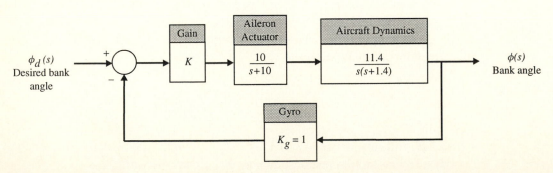

FIGURE 5.3
Bank angle control autopilot.

5.3   BANK ANGLE CONTROL DESIGN

The closed-loop transfer function is

$$T(s) = \frac{\phi(s)}{\phi_d(s)} = \frac{114K}{s^3 + 11.4s^2 + 14s + 114K} \;.$$ (5.5)

We want to determine analytically the values of K that will give us the desired response—namely, a percent overshoot less than 20% and a fast time-to-peak. The analytic analysis would be simpler if our closed-loop system were a second-order system (since we have valuable relationships between settling time, percent overshoot, natural frequency and damping ratio); however we have a third-order system, given by $T(s)$ in Eq. (5.5). We could consider approximating the third-order transfer function by a second-order transfer function—this is sometimes a very good engineering approach to analysis. There are many methods available to obtain approximate transfer functions. Here we use an algebraic method (described in Chapter 5 of *Modern Control Systems*) that attempts to match the frequency response of the approximate system as closely as possible to the actual system.

Our transfer function can be rewritten as

$$T(s) = \frac{1}{1 + \frac{14}{114K}s + \frac{11.4}{114K}s^2 + \frac{1}{114K}s^3} \;,$$

by factoring the constant term out of the numerator and denominator. Suppose our approximate transfer function is given by the second-order system

$$L(s) = \frac{1}{1 + d_1 s + d_2 s^2} \;.$$

The objective is to find appropriate values of d_1 and d_2. We define

$$M^{(k)}(s) \triangleq \frac{d^k}{ds^k} M(s) \;,$$ (5.6)

and

$$\Delta^{(k)}(s) \triangleq \frac{d^k}{ds^k} \Delta(s) \;,$$ (5.7)

where $M(s)$ is the denominator polynomial of $L(s)$, and $\Delta(s)$ is the denominator polynomial of $T(s)$. We also define

$$M_{2q} = \sum_{k=0}^{2q} \frac{(-1)^{(k+q)} M^{(k)}(0) M^{(2q-k)}(0)}{k!(2q-k)!} \;, \qquad q = 1, 2, \ldots \;,$$ (5.8)

and

$$\Delta_{2q} = \sum_{k=0}^{2q} \frac{(-1)^{(k+q)} \Delta^{(k)}(0) \Delta^{(2q-k)}(0)}{k!(2q-k)!}, \quad q = 1, 2, \ldots . \qquad (5.9)$$

Then, forming the set of algebraic equations

$$M_{2q} = \Delta_{2q}, \quad q = 1, 2, \ldots, \qquad (5.10)$$

we can solve for the unknown parameters of the approximate function. The index q is incremented until sufficient equations are obtained to solve for the unknown coefficients of the approximate function. In this case, $q = 1, 2$ since we have two parameters d_1 and d_2 to obtain. We have

$$M(s) = 1 + d_1 s + d_2 s^2 .$$

So from Eq. (5.6) it follows that

$$M^{(1)}(s) = \frac{dM}{ds} = d_1 + 2d_2 s ,$$

$$M^{(2)}(s) = \frac{d^2 M}{ds^2} = 2d_2 ,$$

$$M^{(3)}(s) = M^4(s) = \ldots = 0 .$$

Thus evaluating at $s = 0$ yields

$$M^{(1)}(0) = d_1 ,$$

$$M^{(2)}(0) = 2d_2 ,$$

$$M^{(3)}(0) = M^{(4)}(0) = \ldots = 0 .$$

Similarly, with $\Delta(s)$ defined as the denominator of the closed-loop transfer function $T(s)$, we have

$$\Delta(s) = 1 + \frac{14}{114K} s + \frac{11.4}{114K} s^2 + \frac{s^3}{114K} .$$

Using Eq. (5.7) yields

$$\Delta^{(1)}(s) = \frac{d\Delta}{ds} = \frac{14}{114K} + \frac{22.8}{114K} s + \frac{3}{114K} s^2 ,$$

$$\Delta^{(2)}(s) = \frac{d^2 \Delta}{ds^2} = \frac{22.8}{114K} + \frac{6}{114K} s ,$$

$$\Delta^{(3)}(s) = \frac{d^3 \Delta}{ds^3} = \frac{6}{114K} ,$$

$$\Delta^{(4)}(s) = \Delta^5(s) = \ldots = 0 .$$

Evaluating at $s = 0$, it follows that

$$\Delta^{(1)}(0) = \frac{14}{114K} \, ,$$

$$\Delta^{(2)}(0) = \frac{22.8}{114K} \, ,$$

$$\Delta^{(3)}(0) = \frac{6}{114K} \, ,$$

$$\Delta^{(4)}(0) = \Delta^5(0) = \ldots = 0 \, .$$

Using Eq. (5.8) for $q = 1$ and $q = 2$ yields

$$M_2 = -\frac{M^o(0)M^2(0)}{2} + \frac{M^1(0)M^1(0)}{1} - \frac{M^2(0)M^o(0)}{2} = -2d_2 + d_1^2 \, ,$$

and

$$M_4 = \frac{(-1)^2 M^o(0)M^1(0)}{0!4!} + \frac{(-1)^3 M^1(0)M^3(0)}{1!3!} + \frac{(-1)^4 M^2(0)M^2(0)}{2!2!}$$

$$+ \frac{(-1)^5 M^3(0)M^1(0)}{3!1!} + \frac{(-1)^6 M^4(0)M^o(0)}{4!0!} = d_2^2 \, .$$

Similarly using Eq. (5.9), we find that

$$\Delta_2 = \frac{-22.8}{114K} + \frac{196}{(114K)^2} \, ,$$

and

$$\Delta_4 = \frac{101.96}{(114K)^2} \, .$$

Thus forming the set of algebraic equations in Eq. (5.10), we obtain

$$M_2 = \Delta_2$$
$$M_4 = \Delta_4 \, ,$$

or

$$-2d_2 + d_1^2 = \frac{-22.8}{114K} + \frac{196}{(114K)^2} \, ,$$

and

$$d_2^2 = \frac{101.96}{(114K)^2} \, .$$

Solving for d_1 and d_2 yields

$$d_1 = \frac{\sqrt{196 - 296.96K}}{114K} , \tag{5.11}$$

$$d_2 = \frac{10.097}{114K} , \tag{5.12}$$

where we always choose the positive values of d_1 and d_2 so that $L(s)$ has poles in the left half-plane. Thus (after some manipulation) the approximate transfer function is

$$L(s) = \frac{11.29K}{s^2 + \sqrt{1.92 - 2.91K}\,s + 11.29K} . \tag{5.13}$$

We require that $K < 0.65$ so that the coefficient of the s term remains a real number (we do not want to have a transfer function with complex valued parameters).

Our desired second-order transfer function can be written as

$$L(s) = \frac{\omega_n^2}{s^2 + 2\zeta\omega_n + \omega_n^2} . \tag{5.14}$$

Comparing coefficients in Eqs. (5.13) and (5.14) yields

$$\omega_n^2 = 11.29K ,$$

$$\zeta^2 = \frac{0.043}{K} - 0.065 . \tag{5.15}$$

The design specification of percent overshoot, $P.O.$, less than 20% implies that we want $\zeta \geq 0.45$. This follows from solving

$$P.O. = 100e^{\frac{-\pi\zeta}{\sqrt{1-\zeta^2}}} \leq 20$$

for ζ. Setting $\zeta = 0.45$ in Eq. (5.15) and solving for K yields

$$K = 0.16 .$$

With $K = 0.16$ we compute

$$\omega_n = \sqrt{11.29K} = 1.34 .$$

Then we can estimate the time-to-peak, T_p, to be

$$T_p = \frac{\pi}{\omega_n\sqrt{1 - \zeta^2}} = 2.62 \text{ seconds} .$$

We might be tempted at this point to select $\zeta > 0.45$ so that we reduce the percent overshoot even further than 20%. What happens if we decide to try this approach? From Eq. (5.15) we see that K decreases as ζ increases. Then, since

$$\omega_n = \sqrt{11.29K} \, ,$$

as K decreases, then ω_n also decreases. But the time-to-peak, given by

$$T_p = \frac{\pi}{\omega_n\sqrt{1 - \zeta^2}} \, ,$$

increases as ω_n decreases. Since our goal is to meet the specification of percent overshoot less than 20% while minimizing the time-to-peak, we use the initial selection of $\zeta = 0.45$ so that we do not increase T_p unnecessarily.

The second-order system approximation has allowed us to gain insight into the relationship between the parameter K and the system response, as measured by percent overshoot and time-to-peak. Of course, the gain $K = 0.16$ is only a starting point in the design. Why? Because we in fact have a third-order system and must consider the effect of the third pole (which we have ignored so far).

5.4   SIMULATION DEVELOPMENT

We will use SIMULINK to develop a simulation of the control system. We quantify the performance using the initial value of $K = 0.16$. If necessary, we can easily vary important system parameters (such as K) and check the resulting performance (as measured by step response characteristics). We need not use SIMULINK to perform the simulation function; using MATLAB scripts can also be very effective. However, SIMULINK provides a graphical methodology for developing the simulation using block diagrams. Therefore we should view SIMULINK as just another tool to assist us in our design and analysis.

As shown in Figure 5.4, SIMULINK presents the window of the system in block diagram form. Before starting the simulation we can choose to change many of the parameters of the system we are modeling. To change any part in the model, we simply open (double-click with the mouse button) the Gain, Aileron actuator, Aircraft dynamics, or Gyro gain box and enter a new parameter.

For example, opening the Gain box displays the window shown in Figure 5.5. We use this window to specify the value of the gain introduced into the system. To move the window and avoid obstructing the diagram, we can click and hold the mouse button anywhere on the title bar of the simulation control window (where the title Bank_Angle_1 appears). As we move the mouse we drag the window some place else. As we change the parameters they are immediately updated on the diagram. Figures 5.6 and 5.7 show the display windows opened for the aileron actuator and the aircraft dynamics.

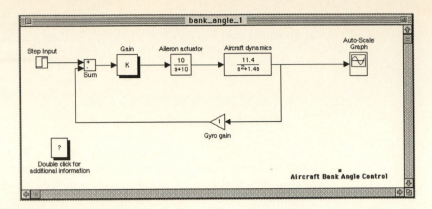

FIGURE 5.4
SIMULINK simulation window with aircraft, controller, and actuator dynamics that has been pre-configured for the user.

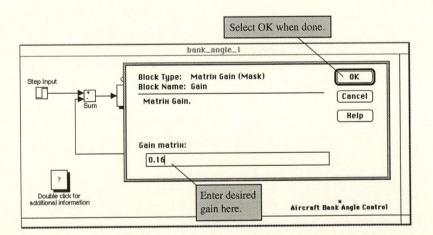

FIGURE 5.5
Pop-up window used to change the scalar value of the gain introduced into the system.

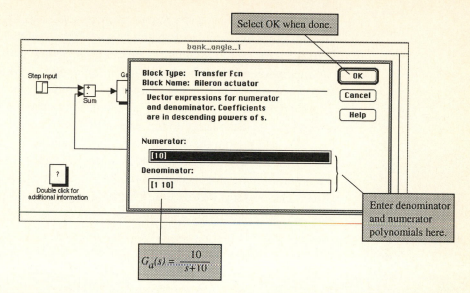

FIGURE 5.6
Pop-up window used to change the numerator and denominator polynomials of the aileron actuator transfer function.

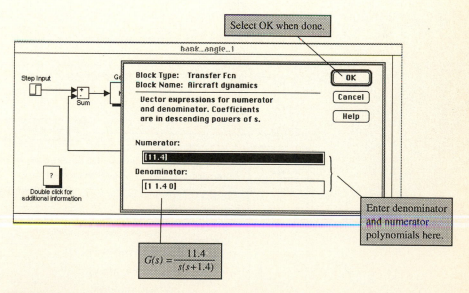

FIGURE 5.7
Pop-up window used to change the numerator and denominator polynomials of the aircraft dynamics transfer function.

We need to set just a few more parameters before the model is ready for simulation. From the Simulation menu choose Parameters, as shown in Figure 5.8. Figure 5.9 shows the Simulation Parameters window with the different numerical integration routines displayed on the left and the parameters for the start and stop time, maximum and minimum step size, relative error, and return variables displayed on the right. After making the desired changes, select the OK button. Always set the Stop Time to the desired value! To begin the simulation, select the Start option from the Simulation menu, as shown in Figure 5.10. The simulation executes until the Stop Time is reached. During the simulation execution, the time response plot (generated by the Auto-Scale Graph) is dynamically updated. The desktop should now resemble Figure 5.11. Your desktop may not appear exactly the same, depending on your computer system, software version,

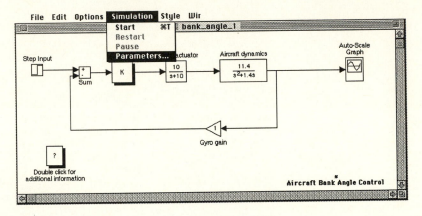

FIGURE 5.8
Selecting Parameters from the Simulation menu.

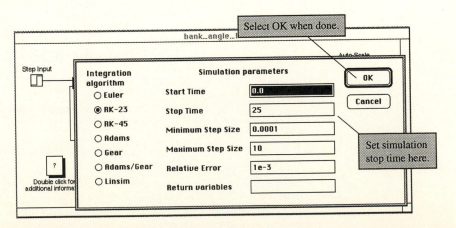

FIGURE 5.9
Setting the various simulation parameters.

and how you have arranged things on the desktop. To fine-tune the simulation, we can change any or all of the transfer functions or gains by the steps that we just covered.

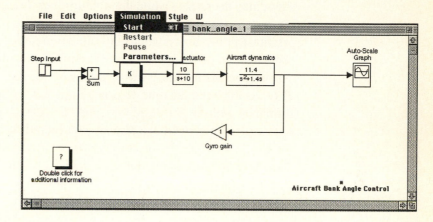

FIGURE 5.10
Starting the simulation.

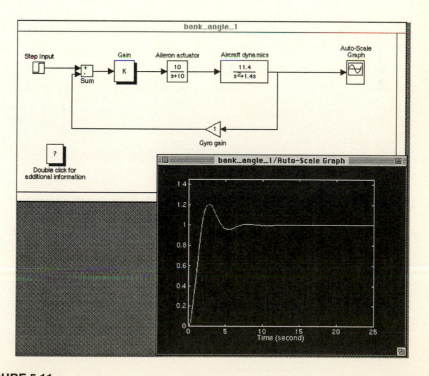

FIGURE 5.11
The SIMULINK simulation with the scaled graph showing the aircraft bank angle response to a step input.

5.4.1 Simulation Analysis

At this point the SIMULINK set-up has $K = 0.16$ (see Figure 5.5). From our previous analytic analysis, we determined that as K decreased, the percent overshoot also decreases, while the time-to-peak simultaneously increases. With an interactive simulation we can verify this trend, except that now we have the original third-order system in the simulation rather than the approximate second-order system on which the analytic analysis was based. With the second-order system approximation, we estimate that with $K = 0.16$ the percent overshoot, $P.O.$, is 20% and the time-to-peak, T_p, is 2.62 seconds. What is the system performance with the third-order system? SIMULINK provides a way to investigate this easily. Since we have already configured the simulation with $K = 0.16$, we can then select the Start option from the Simulation menu. The results of the simulation are shown in Figure 5.11. The percent overshoot is slightly more than 20% due to the presence of a third pole that we ignored in the second-order system analytic analysis. In fact, with $K = 0.16$ we have

$$P.O. = 20.5\% \quad \text{and} \quad T_p = 2.73 \text{ seconds} .$$

It can be rather difficult to determine precisely the overshoot and time-to-peak from the auto-scale graph. To determine percent overshoot and time-to-peak more accurately, we can write certain parameters to the MATLAB workspace. Figure 5.12 shows two additions made to the original SIMULINK diagram so that the aircraft bank angle and simulation time are written to the workspace.

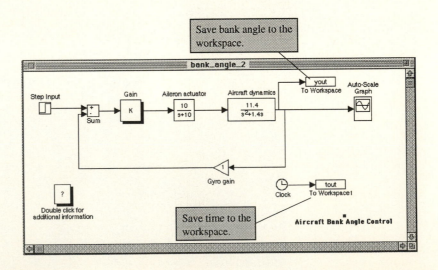

FIGURE 5.12
Enhanced SIMULINK diagram includes writing the aircraft bank angle and simulation time to the workspace so that percent overshoot and time-to-peak can be determined accurately.

When the simulation stops we can use the MATLAB command

$$[\text{ymax},\text{I}]=\max(\text{yout}) ,$$

to determine the peak output from which the percent overshoot can be determined. Also, the integer I contains the index at which the peak value of yout occurs. Thus, the time-to-peak is the I-th element of the tout variable (which now exists in the workspace).

For comparison purposes, select two variations to the gain K:

$$K_1 = 0.1$$
$$K_2 = 0.2 .$$

Follow these steps (as described in the previous section):

1.   Double-click on the Gain box.

2.   In the pop-up window, change the gain from K = 0.16 to K = 0.1.

3.   Click OK to close the pop-up window.

4.   Select the Start option under the Simulation menu.

After the simulation is finished, the Auto-Scale Graph displays the results of the step response, as shown in Figure 5.13. The percent overshoot is reduced from 20.5% (with $K = 0.16$) to 9.5% (with $K = 0.1$). Also the time-to-peak

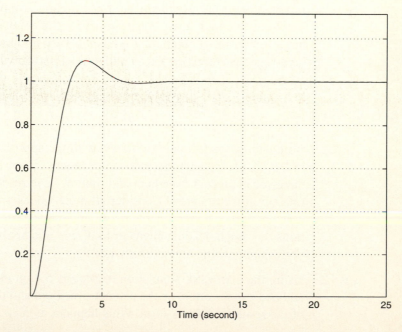

FIGURE 5.13
Simulation results with $K = 0.1$.

increases from 2.73 seconds (with $K = 0.16$) to 3.74 seconds (with $K = 0.1$). We can verify that when $K = 0.2$, the percent overshoot and time-to-peak are 26.5% and 2.38 seconds, respectively. The results are shown in Table 5.1.

TABLE 5.1   Performance Comparison for K = 0.1, 0.16 and 0.2

K	$P.O.$ (%)	T_p (seconds)
0.10	9.5	3.74
0.16	20.5	2.73
0.20	26.5	2.38

5.5   SUMMARY

In this chapter we discussed the use of SIMULINK and a block diagram graphical method to develop a simulation of the bank angle motion of an airplane. Using an algebraic method, we simplified the third-order transfer function describing the bank angle response to an aileron deflection input that matches the frequency response magnitude plots of the simplified transfer function to the actual transfer function. Then we used the resulting simplified model (a second-order transfer function) to obtain a reasonable value for the controller gain K. After producing a SIMULINK simulation, it is easy to vary important system parameters and investigate the effect on the overall system performance.

EXERCISES

E5.1   Compare the performance obtained with the approximate aircraft model, given in Eq. (5.13), against the third-order model, given in Eq. (5.5). How much is the error in percent overshoot and time-to-peak results when using the approximate transfer function? Consider all three values of $K = 0.1$, 0.16, and 0.2. In practice, it may be reasonable to use an approximate model. Discuss the conditions that might lead an engineer to design and analyze the closed-loop system performance using approximate models.

E5.2   Using the SIMULINK simulation in Section 5.4, gradually increase the gain, K, and observe the effect on the percent overshoot and time-to-peak. Keep a record of the results in tabular form. What happens as $K > 1.4$? Discuss the results.

E5.3   Design a PID controller for the aircraft bank angle control problem presented in this chapter to meet the same performance specifications (percent overshoot less than 20% and minimum time-to-peak). A PID controller has the form

$$G_c(s) = K_p + \frac{K_I}{s} + K_D s \ .$$

What additional performance can be obtained with the PID controller? Would you recommend a PID controller over a proportional controller? Discuss.

Stability of Linear Feedback Systems

Robot-controlled Motorcycle Example

6.1 Introduction 105

6.2 BIBO Stability 105

6.3 Robot-controlled Motorcycle 111

6.4 Stability Analysis 114

6.5 Disturbance Response 116

6.6 MATLAB Analysis 117

6.7 Summary 120

Exercises 120

P R E V I E W

The main topic of this chapter is stability. In particular the notion of bounded-input bounded-output (BIBO) stability is discussed. We use a simple mass-spring-damper system to make the point that some systems are inherently stable and that we can establish BIBO stability by showing that the integrated absolute value of the impulse response is finite. Equivalently, a system is stable if all the roots of the characteristic equation (that is, all the poles) lie in the left half-plane. Finally, we use the problem of a robot-controlled motorcycle to illustrate the role of the Routh-Hurwitz method in establishing stability and aiding in the selection of key system tuning parameters.

6.1   INTRODUCTION

Of utmost importance in the design of control systems is stability. In some cases the open-loop system can be unstable; our objective is to stabilize the system using closed-loop control while realizing certain performance objectives. In other cases the open-loop system is inherently stable, and we want to improve the transient response. However we do not want to introduce instability into the closed-loop system while attempting to meet the performance design specifications. Introducing instability into an otherwise stable system is a distinct possibility with control system design. In Chapter 4 of *Modern Control Systems*, we discuss the cost of feedback. Introducing instability is a potential cost of feedback, although the benefits outweigh the cost to such an extent that we find feedback control systems in use in most aspects of our everyday lives (see Chapter 4 in *Modern Control Systems* for a summary of the reasons for using feedback).

In this chapter we use the Routh-Hurwitz method to look at the stability issues of control design, using the problem of a robot-controlled motorcycle. The design elements emphasized in this chapter are illustrated in Figure 6.1. We will concentrate on the selection of key parameters to be adjusted.

6.2   BIBO STABILITY

There are many valid definitions of **system stability**, and there are many factors that affect stability. We say that a system is stable if the system response is bounded for every bounded input. This is sometimes referred to as bounded-input bounded-output or **BIBO stability**. We can also consider open-loop stability versus closed-loop stability. Many systems are inherently stable. For example, consider the mass-spring-damper system shown in Figure 6.2.

The equation of motion for mass m is

$$m\ddot{y} + b\dot{y} + ky = u \, ,$$

where y is the position of mass m, b is the damping coefficient, k is the spring constant, and u is the external force. The corresponding transfer function model is

$$\frac{Y(s)}{U(s)} = \frac{\frac{1}{m}}{s^2 + \frac{b}{m}s + \frac{k}{m}} \, . \tag{6.1}$$

A typical input response is shown in Figure 6.3, where $m = 10$ slugs, $k = 1$ lb/ft, $b = 2$ lb-sec/ft, and the step input is given by $u(t) = 1$ lb. In this case the input is bounded in magnitude because $|u(t)| = 1$, and the output (shown in Figure 6.3) is also bounded in magnitude because $|y(t)| \leq 1.35$. Since we have a second-order system, we can easily compute the maximum value of the output to a unit

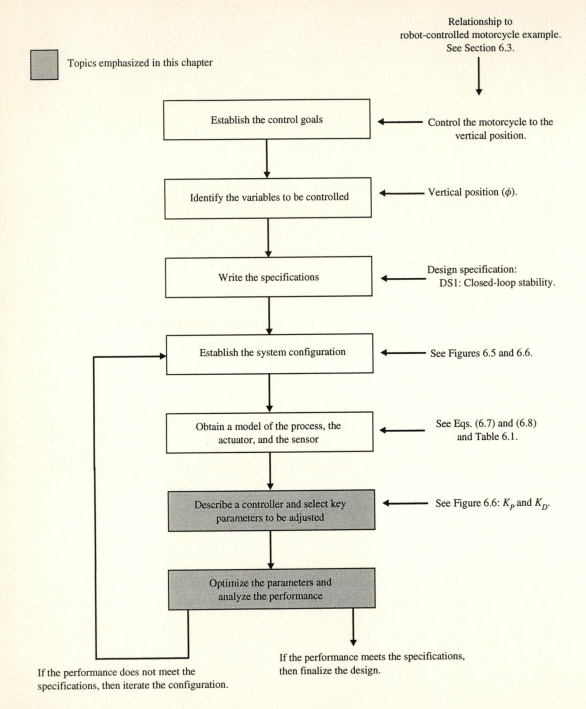

□ Topics emphasized in this chapter

Relationship to robot-controlled motorcycle example. See Section 6.3.

Establish the control goals ⟵ Control the motorcycle to the vertical position.

Identify the variables to be controlled ⟵ Vertical position (ϕ).

Write the specifications ⟵ Design specification: DS1: Closed-loop stability.

Establish the system configuration ⟵ See Figures 6.5 and 6.6.

Obtain a model of the process, the actuator, and the sensor ⟵ See Eqs. (6.7) and (6.8) and Table 6.1.

Describe a controller and select key parameters to be adjusted ⟵ See Figure 6.6: K_P and K_D.

Optimize the parameters and analyze the performance

If the performance does not meet the specifications, then iterate the configuration.

If the performance meets the specifications, then finalize the design.

FIGURE 6.1
Elements of the control system design process emphasized in this chapter.

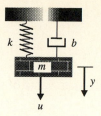

FIGURE 6.2
The mass-spring-damper system is inherently stable.

```
m=10; b=2; k=1;
num=[1/m]; den=[1 b/m k/m];
[y,x,t]=step(num,den);
plot(t,y), grid
xlabel('Time (secs)'),ylabel('y (ft)')
ymax=max(y)
```

ymax =
1.3507

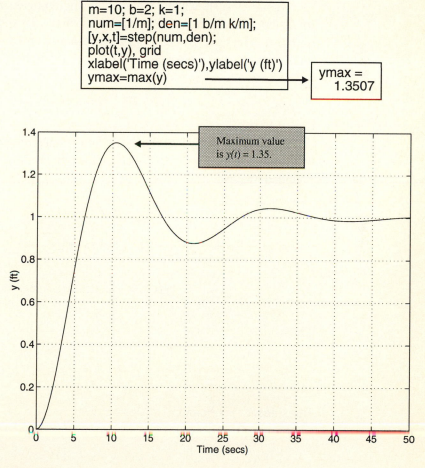

FIGURE 6.3
Unit step response for the mass-spring-damper system.

step input. Using the parameter values $m = 10$, $k = 1$, $b = 2$, we determine that

$$\omega_n = \sqrt{k/m} = \sqrt{0.1} = 0.316 \quad \text{and} \quad \zeta = \frac{b}{2\omega_n m} = 0.316 \, .$$

Computing the percent overshoot yields

$$P.O. = 100e^{-\pi\zeta/\sqrt{1-\zeta^2}} = 35\% \ .$$

The steady-state value of the step response is

$$\lim_{t\to\infty} y(t) = \lim_{s\to 0} s\frac{\frac{1}{m}}{s^2 + \frac{b}{m}s + \frac{k}{m}}\frac{1}{s} = \frac{1}{k} = 1 \ ,$$

where $k = 1$. A 35% overshoot corresponds to a maximum value of 1.35.

We have just shown that the mass-spring-damper system response is bounded for a step input. How do we know that the system response is bounded for every bounded input? Let $h(t)$ represent the response of a system to an impulse input. Then the output of the system, $y(t)$, to any input, $u(t)$, is given by

$$y(t) = \int_0^t h(t)u(t - \tau)d\tau \ . \tag{6.2}$$

Eq. (6.2) is known as the convolution integral. So in the time-domain the input-output behavior is described by the convolution. Suppose that the input is bounded, that is

$$|u(t)| \le M,$$

where $M < \infty$. From Eq. (6.2) it follows that

$$|y(t)| = \left|\int_0^t h(t)u(t - \tau)d\tau\right| \ .$$

Thus

$$|y(t)| \le \int_0^t |h(t)||u(t - \tau)|d\tau \ .$$

But since $u(t)$ is bounded by M, we have

$$|y(t)| \le M\int_0^t |h(t)|d\tau \ .$$

If $\int_0^t |h(\tau)|d\tau$ is bounded by a finite number, then it follows that $y(t)$ is bounded. It can also be shown that $y(t)$ being bounded implies $\int_0^t |h(\tau)|d\tau$ is bounded (for bounded $u(t)$).

To determine whether the mass-spring-damper system is stable for bounded inputs, we need only to check that $\int_0^t |h(\tau)|d\tau$ is bounded, where $h(t)$ is the impulse response function. The response of the system to an input $U(s)$ is

$$Y(s) = \frac{1}{k} \frac{\frac{k}{m}}{s^2 + \frac{b}{m}s + \frac{k}{m}} U(s) . \tag{6.3}$$

We can write a general second-order system as

$$Y(s) = \frac{\omega_n^2}{s^2 + 2\zeta\omega_n s + \omega_n^2} U(s) . \tag{6.4}$$

Comparing Eqs. (6.3) and (6.4), we establish the relationships

$$\omega_n^2 = k/m ,$$
$$2\zeta\omega_n = b/m .$$

Then using the Laplace transform tables (see *Modern Control Systems*, Appendix A) with $U(s) = 1$ (that is, an impulse) we obtain

$$y(t) = \frac{1}{\sqrt{mk - (b/2)^2}} \; e^{-\frac{b}{2m}t} \sin\frac{1}{m}\sqrt{mk - (b/2)^2t} ,$$

which is valid when $\zeta < 1$, or in our case when

$$\frac{b}{2\sqrt{km}} < 1 . \tag{6.5}$$

Also $h(t) = y(t)$, since we defined $h(t)$ to be the system response to an impulse input. We compute

$$|h(t)| = \left| \frac{1}{\sqrt{mk - (b/2)^2}} e^{-(1/m)(b/2)t} \sin\frac{1}{m}\sqrt{mk - (b/2)^2t} \right|$$

$$\leq \frac{1}{\sqrt{mk - (b/2)^2}} e^{-(1/m)(b/2)t} .$$

If we suppose that the impulse is applied at $t = 0$ and that the system is causal (there is no output before the input is applied), then

$$|h(t)| \leq \frac{1}{\sqrt{mk - (b/2)^2}} .$$

Since $b < 2\sqrt{km}$ [see Eq. (6.5)], the impulse response is bounded. Therefore, the system output is bounded for every bounded input. We do not have to check the impulse response every time to verify BIBO stability.

A necessary and sufficient condition for a system to be BIBO stable is that all of the poles of the system transfer function have negative real parts. The poles of the mass-spring-damper system [see Eq. (6.1)] are

$$s = -\zeta\omega_n \pm \omega_n\sqrt{\zeta^2 - 1} \,.$$

For $0 < \zeta < 1$ and $\omega_n > 0$, the poles have negative real parts, and the mass-spring-damper system is BIBO stable. Thus we can conclude that the system is inherently stable.

The location of the poles of the system transfer function are critical to stability. It is well known that the pole locations also affect the system performance, as measured by overshoot, settling time, time-to-peak, and so on (see Chapters 4 and 5 in *Modern Control Systems*). Knowledge of the specific location of the poles within the left half-plane is not necessary for stability considerations. We need only to know that all the poles lie in the left half-plane. If any pole is in the right half-plane, then the system is said to be unstable. If any simple poles lie on the $j\omega$ axis, the system is marginally stable. If any repeated poles lie on the $j\omega$ axis, the system is unstable. For example a system with two poles at $s_{1,2} = \pm 2j$ is marginally stable, while a system with poles at $s_{1,2} = \pm 2j$ and $s_{3,4} = \pm 2j$ is unstable.

The **Routh-Hurwitz method** is a way of determining how many roots of a polynomial have positive real parts (that is, how many lie in the right half-plane). In control applications the polynomial of interest is the characteristic polynomial. The characteristic polynomial of the mass-spring-damper system [see Eq. (6.1)] is

$$q(s) = s^2 + \frac{b}{m}s + \frac{k}{m} = 0 \,.$$

Since hand calculators compute zeros of polynomials quickly and accurately, why not just compute numerically the zeros of the characteristic polynomial? To answer this question, consider the feedback system shown in Figure 6.4.

The closed-loop transfer function is

$$T(s) = \frac{K(s^2 + 2s + 2)}{(K + 1)s^2 + (5 + 2K)s + (6 + 2K)} \,.$$

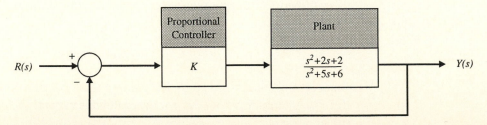

FIGURE 6.4
A simple feedback control system with proportional controller gain K.

The corresponding characteristic polynomial is

$$q(s) = (K + 1)s^2 + (5 + 2K)s + (6 + 2K) = 0 \,. \qquad (6.6)$$

For what values of K is the closed-loop system stable? We can answer this question using an elegant method, called the root locus method, which is the subject of the next chapter. A brute-force method for answering that question is to select a value of K, use a calculator to compute the zeros of the characteristic equation, and check to see if any lie in the right half-plane. This process is then repeated until we are confident that we have discovered all Ks for which the system is stable; this is neither an elegant nor a practical approach. Fortunately, the Routh-Hurwitz criterion provides a straightforward way to answer the stability question. As a design tool, we can use the Routh-Hurwitz criterion to establish valid values of the controller gain K so that stability is guaranteed.

The Routh array for the characteristic polynomial in Eq. (6.6) is

$$
\begin{array}{c|cc}
s^2 & K + 1 & 6 + 2K \\
s & 5 + 2K & 0 \\
1 & 6 + 2K &
\end{array}
$$

The Routh-Hurwitz criterion states that for closed-loop stability, all terms in the first column of the Routh array must be positive. Thus we have

$$
\begin{aligned}
K + 1 > 0 &\quad\Rightarrow\quad K > -1 \,, \\
5 + 2K > 0 &\quad\Rightarrow\quad K > -2.5 \,, \\
6 + 2K > 0 &\quad\Rightarrow\quad K > -3 \,.
\end{aligned}
$$

Choosing $K > -1$ satisfies all three inequalities, thus implying that the closed-loop system is stable (that is, no poles of the closed-loop system are in the right half-plane). In the design process we can select any $K > -1$ to meet the performance specifications (such as percent overshoot, settling time, and so on) and know that the closed-loop system will remain stable.

6.3   ROBOT-CONTROLLED MOTORCYCLE

Consider the robot-controlled motorcycle shown in Figure 6.5. This problem is adapted from DP6.6 in *Modern Control Systems*. The motorcycle will move in a straight line at constant forward speed V. Let ϕ denote the angle between the plane of symmetry of the motorcycle and the vertical. The desired angle ϕ_d is equal to zero:

$$\phi_d(s) = 0 \,.$$

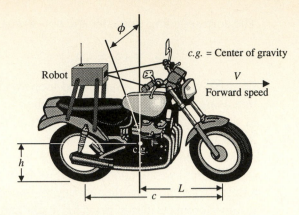

FIGURE 6.5
The robot-controlled motorcyle.

The control goal is

Control Goal   Control the motorcycle in the vertical position, and maintain the prescribed position in the presence of disturbances.

The variable to be controlled is

Variable to Be Controlled   The motorcycle position from vertical (ϕ).

Since our focus here is on stability rather than transient response characteristics, the control specifications will be related to stability only; transient performance is an issue that we need to address once we have investigated all the stability issues. The control design specification is

Design Specification   **DS1** The closed-loop system must be stable.

The main components of the robot-controlled motorcycle are the motorcycle and robot, the controller and the feedback measurements. The main subject of the chapter is not modeling, so we do not concentrate on developing the motorcycle dynamics model. We rely instead on the work of others (see [52]). The motorcycle model is given by

$$G(s) = \frac{1}{s^2 - \alpha_1}, \qquad (6.7)$$

where $\alpha_1 = g/h$, $g = 9.806$ m/sec^2, and h is the height of the motorcycle center of gravity above the ground (see Figure 6.5). The motorcycle is unstable with poles at $s = \pm\sqrt{\alpha_1}$. The controller is given by

$$G_c(s) = \frac{\alpha_2 + \alpha_3 s}{\tau s + 1} , \tag{6.8}$$

where

$$\alpha_2 = V^2/(hc)$$

and

$$\alpha_3 = VL/(hc) .$$

The forward speed of the motorcycle is denoted by V, and c denotes the wheel-base (the distance between the wheel centers). The length, L, is the horizontal distance between the front wheel axle and the motorcycle center of gravity. The time-constant of the controller is denoted by τ. This term represents the speed of response of the controller; smaller values of τ indicate an increased speed of response. Many simplifying assumptions are necessary to obtain the simple transfer function models in Eqs. (6.7) and (6.8).

Control is accomplished by turning the handlebar. The front wheel rotation about the vertical is not evident in the transfer functions. Also, the transfer functions assume a constant forward speed, V, which means that we must have another control system at work regulating the forward speed. Nominal motorcycle and robot controller parameters are given in Table 6.1.

Assembling the components of the feedback system gives us the system configuration shown in Figure 6.6. Examination of the configuration reveals that the robot controller block is a function of the physical system (h, c, and L), the operating conditions (V), and the robot time-constant (τ). No parameters need ad-

TABLE 6.1   Physical Parameters

τ	0.2	sec
α_1	9	1/sec^2
α_2	2.7	1/sec^2
α_3	1.35	1/sec
h	1.09	m
V	2.0	m/sec
L	1.0	m
c	1.36	m

justment unless we physically change the motorcycle parameters and/or speed. In fact, in this example the parameters we want to adjust are in the feedback loop:

Select Key Tuning Parameters

Feedback gains K_P and K_D.

The key tuning parameters are not always in the forward path; in fact they may exist in any subsystem in the block diagram.

We want to use the Routh-Hurwitz technique to analyze the closed-loop system stability. What values of K_P and K_D lead to closed-loop stability? A related question that we can pose is, given specific values of K_P and K_D for the nominal system (that is, nominal values of $\alpha_1, \alpha_2, \alpha_3$, and τ), how can the parameters themselves vary while still retaining closed-loop stability?

6.4 STABILITY ANALYSIS

The closed-loop transfer function from $\phi_d(s)$ to $\phi(s)$ is

$$T(s) = \frac{\alpha_2 + \alpha_3 s}{\Delta(s)} ,$$

where

$$\Delta(s) = \tau s^3 + (1 + K_D \alpha_3)s^2 + (K_D \alpha_2 + K_P \alpha_3 - \tau \alpha_1)s + K_P \alpha_2 - \alpha_1 .$$

The characteristic equation is

$$\Delta(s) = 0 .$$

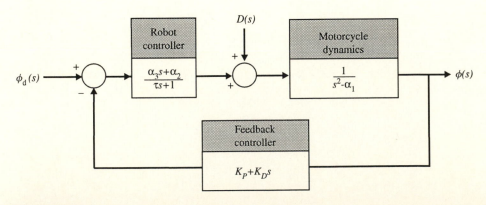

FIGURE 6.6
The robot-controlled motorcyle feedback system block diagram.

The question that we need to answer is for what values of K_P and K_D does the characteristic equation $\Delta(s) = 0$ have all zeros in the left half-plane?

We can set up the following Routh array:

$$
\begin{array}{c|cc}
s^3 & \tau & K_D\alpha_2 + K_P\alpha_3 - \tau\alpha_1 \\
s^2 & 1 + K_D\alpha_3 & K_P\alpha_2 - \alpha_1 \\
s & a & \\
1 & K_P\alpha_2 - \alpha_1 &
\end{array}
$$

where

$$
a = \frac{(1 + K_D\alpha_3)(K_D\alpha_2 + K_P\alpha_3 - \tau\alpha_1) - \tau(\alpha_2 K_P - \alpha_1)}{1 + K_D\alpha_3}.
$$

By inspecting column 1, we determine that for stability we require

$$
\tau > 0 ,
$$
$$
K_D > -1/\alpha_3 ,
$$
$$
K_P > \alpha_1/\alpha_2 ,
$$
$$
a > 0 .
$$

Choosing $K_D > 0$ satisfies the second inequality (note that $\alpha_3 > 0$). In the event $\tau = 0$, we would reformulate the characteristic equation and rework the Routh array.

The computational difficulty arises in determining the conditions on K_P and K_D such that $a > 0$. Using the nominal values of the parameters α_1, α_2, α_3, and τ (see Table 6.1) and given that $\tau > 0$, $K_D > 0$, and $K_P > \alpha_1/\alpha_3 = 3.33$, we find that $a > 0$ implies that the following relationship must be satisfied:

$$
f = \alpha_2\alpha_3 K_D^2 + (\alpha_2 - \tau\alpha_1\alpha_3 + \alpha_3^2 K_P)K_D + (\alpha_3 - \tau\alpha_2)K_P > 0 . \tag{6.9}
$$

The stability region is shown in Figure 6.7. Taking into account all the inequalities, a valid region for selecting the gains is $K_D > 0$ and $K_P > 3.33$.

Selecting any point (K_P, K_D) in the stability region yields a valid (that is, stable) set of gains for the feedback loop. For example, selecting

$$
K_P = 10 ,
$$
$$
K_D = 5 ,
$$

yields a stable closed-loop system. The closed-loop poles (see Figure 6.8) are

$$
s_1 = -35.2477 ,
$$
$$
s_2 = -2.4674 ,
$$
$$
s_3 = -1.0348 .
$$

```
a1=9;a2=2.7;a3=1.35;tau=0.2;
[KP,KD] = meshgrid(0:1:50, 0:1:50);
f = a2*a3*KD.^2 + (a2-tau*a1*a3+a3^2*KP) .* KD + (a3-tau*a2)*KP;
meshz(f)
xlabel('KP'), ylabel('KD'), zlabel('f')
grid
```

Eq. (6.9)

meshz generates the 3-D plot.

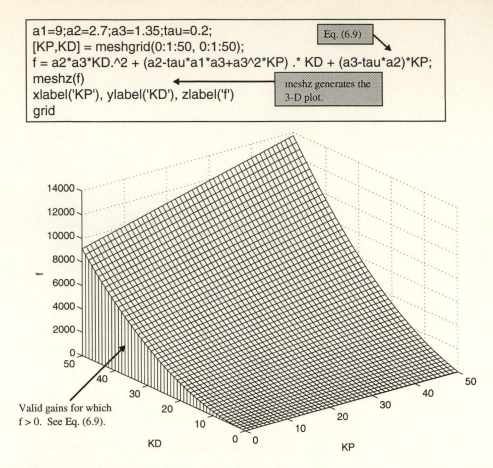

Valid gains for which
f > 0. See Eq. (6.9).

FIGURE 6.7
Region of valid gains (K_D, K_P) for which the inequality in Eq. (6.9) is satisfied.

Since all the poles have negative real parts, we know the system response to any bounded input will be bounded.

For this robot-controlled motorcycle, we do not expect to have to respond to nonzero command inputs (that is, $\phi_d \neq 0$) since we want the motorcycle to remain upright, and we certainly want to remain upright in the presence of external disturbances.

6.5 DISTURBANCE RESPONSE

The transfer function for the disturbance $D(s)$ to the output $\phi(s)$ without feedback is

$$\phi(s) = \frac{1}{s^2 - \alpha_1} D(s).$$

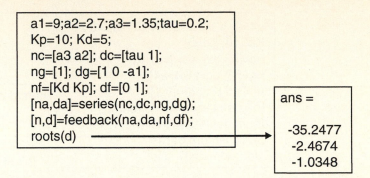

```
a1=9;a2=2.7;a3=1.35;tau=0.2;
Kp=10; Kd=5;
nc=[a3 a2]; dc=[tau 1];
ng=[1]; dg=[1 0 -a1];
nf=[Kd Kp]; df=[0 1];
[na,da]=series(nc,dc,ng,dg);
[n,d]=feedback(na,da,nf,df);
roots(d)
```

```
ans =

 -35.2477
  -2.4674
  -1.0348
```

FIGURE 6.8
Roots of the characteristic equation with $K_P = 10$ and $K_D = 5$.

The characteristic equation is

$$q(s) = s^2 - \alpha_1 = 0 \ .$$

The system poles are

$$s_1 = -\sqrt{\alpha_1} \ ,$$
$$s_2 = +\sqrt{\alpha_1} \ .$$

Thus we see that the motorcycle is unstable; it possesses a pole in the right half-plane. Without feedback control, any external disturbance will result in the motorcycle falling over. Clearly the need for a control system (usually provided by the human rider) is necessary. With the feedback and robot controller in the loop, the closed-loop transfer function from the disturbance to the output is

$$\frac{\phi(s)}{D(s)} = \frac{\tau s + 1}{\tau s^3 + (1 + K_D\alpha_3)s^2 + (K_D\alpha_2 + K_P\alpha_3 - \tau\alpha_1)s + K_P\alpha_2 - \alpha_1} \ .$$

The response to a step disturbance

$$D(s) = \frac{1}{s} \ ,$$

is shown in Figure 6.9; the response is stable. The control system manages to keep the motorcycle upright, although it is tilted at about $\phi = 0.055$ rad $= 3.18$ deg.

6.6   MATLAB ANALYSIS

It is important to give the robot the ability to control the motorcycle over a wide range of forward speeds. Is it possible for the robot, with the feedback gains as selected ($K_P = 10$ and $K_D = 5$), to control the motorcycle as the

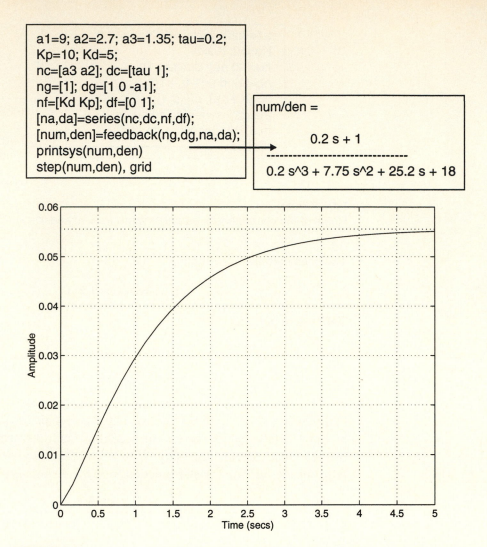

```
a1=9; a2=2.7; a3=1.35; tau=0.2;
Kp=10; Kd=5;
nc=[a3 a2]; dc=[tau 1];
ng=[1]; dg=[1 0 -a1];
nf=[Kd Kp]; df=[0 1];
[na,da]=series(nc,dc,nf,df);
[num,den]=feedback(ng,dg,na,da);
printsys(num,den)
step(num,den), grid
```

```
num/den =

            0.2 s + 1
    ------------------------------
    0.2 s^3 + 7.75 s^2 + 25.2 s + 18
```

FIGURE 6.9
Disturbance response with $K_P = 10$ and $K_D = 5$.

velocity varies? From experience you may know that at slower speeds a bicycle becomes more difficult to control. We expect to see the same characteristics in the stability analysis of our system. Remember that whenever possible, we should try to relate the engineering problem at hand to real-life experiences. This helps to develop intuition that can be used as a reasonableness check on our solution.

A plot of the roots of the characteristic equation as the forward speed, V, varies is shown in Figure 6.10. The data in the plot were generated using the nominal values of the feedback gains, $K_P = 10$ and $K_D = 5$. We selected these gains for the case where $V = 2$ m/sec. Figure 6.10 shows that as V increases, the roots of the characteristic equation remain stable (that is, in the left half-plane) with all points negative. But as the motorcyle forward speed decreases, the roots move toward zero, with one root becoming positive at $V = 1.15$ m/sec. At the point where one root is positive, the motorcycle is unstable.

```
L=1; c=1.36; h=1.09; g=9.806;
v=[0:0.1:5];
a1=g/h;a2=v.^2/h/c;a3=v*L/h/c;tau=0.2;
Kp=10; Kd=5;
ng=[1]; dg=[1 0 -a1];
nf=[Kd Kp]; df=[0 1];
for i=1:length(v)
 nc=[a3(i) a2(i)]; dc=[tau 1];
 [na,da]=series(nc,dc,ng,dg);
 [n,d]=feedback(na,da,nf,df);
r(:,i)=roots(d);
end
plot(v,r,'o'), grid
xlabel('Velocity (m/sec)'), ylabel('Characteristic equation roots')
```

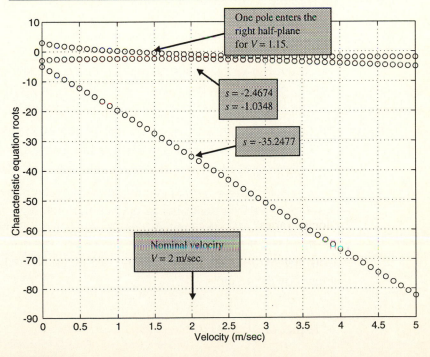

FIGURE 6.10
Roots of the characteristic equation as the motorcycle velocity varies.

6.7   SUMMARY

In this chapter we discussed stability. Using the robot-controlled motorcycle example we showed how to use the Routh-Hurwitz method in the selection of the feedback gains K_P and K_D. Application of the Routh-Hurwitz method resulted in several inequalities that had to be satisfied for the system to be stable. We used these inequalities to develop valid regions for selecting the feedback gains. Given that the robot-controlled motorcycle was stable with gains chosen within the stability region, we studied the motorcycle stability as the motorcycle forward speed varied. We found that as the motorcycle slowed down, the closed-loop system became unstable. We would require a different set of controller gains for the robot-controlled motorcycle at low forward speeds.

E X E R C I S E S

E6.1   To complete the discussion in Section 6.2, show that $y(t)$ being bounded implies $\int_0^t |h(\tau)|d\tau$ is bounded (for bounded $u(t)$).

E6.2   Show that a necessary and sufficient condition for a system to be BIBO stable is that all of the poles of the system transfer function have negative real parts.

E6.3   Suppose the controller in the feedback loop in Figure 6.6 has the form

$$H(s) = \frac{s + p}{s + z} .$$

Use Routh-Hurwitz methods to find the regions of stability for the parameters p and z. Select values for p and z, and analyze the step disturbance response. Evaluate the stability as the motorcycle velocity varies $0 < V \le 5$ m/sec.

E6.4   What changes would you make to the robot-controlled motorcycle (that is, h, c, and L), to make it more robust to changes in forward speed? Interpret robust as an insensitivity of the system stability to changes in forward speed.

E6.5   The task is to develop a robot-controlled motorcycle for operation on the moon. Develop a feedback controller to operate the motorcycle on the moon using a feedback controller structure of your choosing. After showing closed-loop stability to step disturbances on the moon, discuss the use of the same feedback controller on the earth. What realistic disturbances would the motorcycle encounter on the moon?

CHAPTER 7

Root Locus Method

Automobile Velocity Control Example

7.1 Introduction   121

7.2 Sketching a Root Locus   122

7.3 PID Controller   128

7.4 Automobile Velocity Control   130

7.5 Summary   137

Exercises   138

P R E V I E W

In this chapter we discuss the PID controller in the context of the root locus design methodology. Using an actual PID controller design experience as motivation for the root locus method, we discuss the importance of being able to sketch a root locus by hand. Sketching the root locus is a valuable technique in the conceptual design phase of determining valid controller structures (for example, What type of controller will satisfy the design requirements?). We use the automobile velocity control problem to illustrate the root locus method for obtaining a controller that meets our design specifications.

7.1 INTRODUCTION

The **root locus** is a powerful design tool; it is basically a graphical method, and it might not be obvious how we can use it in design. It is essentially a graphical depiction of how the roots of a polynomial change as one parameter, usually denoted by K, in the polynomial varies from $0 < K < \infty$. How can this simple idea be extended to controller design? We look into this in more detail in this chapter. The design elements emphasized in this chapter are depicted in Figure 7.1.

7.2   SKETCHING A ROOT LOCUS

You should know how to sketch a root locus by hand. But as computer-based tools (such as MATLAB) proliferate and enter the mainstream controls community, the question arises over and over again: Why do I need to know how to sketch a root locus by hand when MATLAB will do it for me more accurately?

Some engineers of the slide-rule era compare sketching the root locus by hand in the 1990s with the situation in the 1970s, when the slide rule was being replaced with the hand calculator. There were those who suggested that students should learn how to use a slide rule despite the fact that hand calculators were clearly superior (for example, the calculator took care of the decimal point). Those who supported the continued use of the slide rule lost the argument, and rightly so. The hand calculator (when used properly) could completely replace the slide rule. Engineers simply did not need the slide rule.

The situation with hand sketching of the root locus is not the same. Consider the following example. Suppose we are on a **design team** required to develop a preliminary control system design for a space station. A simple model of the rotation of a space station is

$$G(s) = \frac{1}{Js^2} \, ,$$

where J is the moment of inertia, and at this point in the design process the numerical value is unspecified. The open-loop system is shown in Figure 7.2. Suppose the desired attitude of the space station is $\theta_d = 0^o$ and we want to design the control system to hold attitude. Can this space station be controlled by proportional feedback of the attitude only? The closed-loop feedback system with a contoller, $G_c(s)$, is shown in Figure 7.3.

The closed-loop system characteristic equation is

$$1 + \frac{G_c(s)}{Js^2} = 0 \, .$$

The pole-zero map of the open-loop system is shown in Figure 7.4. Notice that we cannot discern the two poles at the origin using the MATLAB function pzmap. Since the spacecraft parameter J does not affect the location of the open-loop poles, we use a value of $J = 1$ in generating the pole-zero map. With two poles at the origin, the open-loop system is unstable. Without active control the

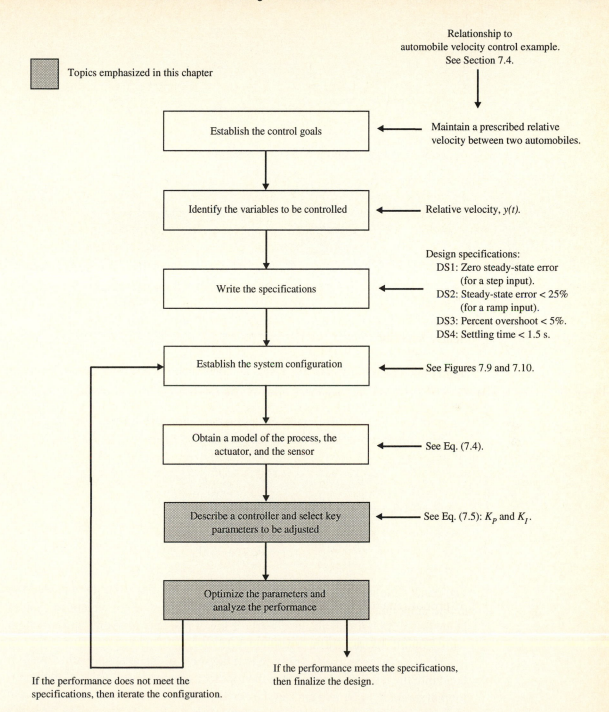

FIGURE 7.1
Elements of the control system design process emphasized in this chapter.

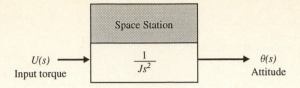

FIGURE 7.2
An open-loop block diagram of a space station.

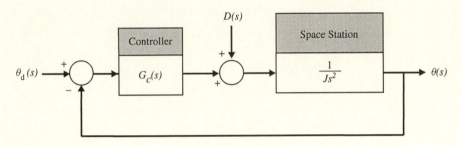

FIGURE 7.3
A closed-loop block diagram of a space station attitude control system.

spacecraft will tumble in the presence of external disturbances.

With proportional feedback of the attitude, we have

$$G_c(s) = K \ ,$$

and the characteristic equation is

$$1 + K\frac{1}{Js^2} = 0 \ .$$

The hand-sketched root locus is shown in Figure 7.5. For stability, we need the loci to enter into the left half-plane. There is no value of K for which the system poles lie in the left half-plane. Clearly it is not possible to stabilize the system with proportional feedback of attitude only.

How can we specify $G_c(s)$ such that the roots of the characteristic equation bend into the left half-plane? Since the root locus begins at the poles of the open-loop system ($K = 0$) and ends at the zeros ($K \to \infty$), it makes sense to locate at least one zero of the controller in the left half-plane to attract the loci. We might consider the controller

$$G_c(s) = K_P + K_D s = K_D \left(s + \frac{K_P}{K_D} \right) \ . \qquad (7.1)$$

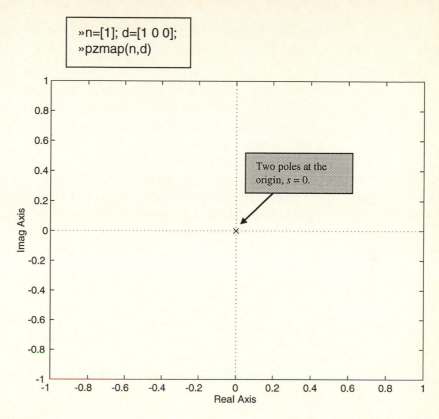

FIGURE 7.4
A pole-zero map of a space station open-loop transfer function.

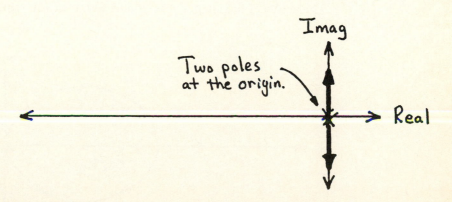

FIGURE 7.5
A hand sketch of the root locus with $G_c(s) = K$ for a space station.

Eq. (7.1) is known as a proportional-derivative (PD) controller. The ratio

$$\gamma = K_P/K_D$$

determines the location of the zero, as shown in Figure 7.6. Stability is not the only issue in design; the **performance specifications** will influence the choice of γ in the final design. The point is that by sketching the root locus, we can quickly answer preliminary questions on controller structure and whether design goals can be met. We see that in the space station case, the attitude rate must also be available for feedback. The feedback control system including the PD controller is shown in Figure 7.7. The PD controller linearly combines the attitude and attitude rate errors. For example with

$$U(s) = G_c(s)E(s) \,,$$

we have

$$u(t) = K_P e(t) + K_D \dot{e}(t) \,.$$

The space station needs to have sensors capable of providing both attitude and attitude rate measurements.

What about a PID (proportional-integral-derivative) controller? Will it also work on a space station? A PID controller is given by

$$G_c(s) = K_P + K_D s + \frac{K_I}{s} = \frac{K_D s^2 + K_P s + K_I}{s} \tag{7.2}$$

$$= K_D \left(\frac{s^2 + \frac{K_P}{K_D}s + \frac{K_I}{K_D}}{s} \right) \,.$$

The action of the integral term improves the steady-state response.

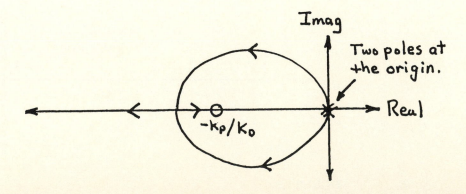

FIGURE 7.6
A hand sketch of the root locus with $G_c(s)$ given in Eq. (7.1) for the space station.

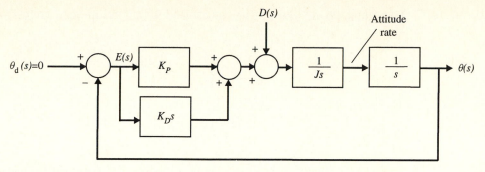

FIGURE 7.7
Closed-loop system with a PD controller.

We see that the PID controller has a pole at $s = 0$ and two zeros at

$$ s = -\frac{K_P}{2K_D} \pm \frac{1}{2}\sqrt{\left(\frac{K_P}{K_D}\right)^2 - 4\frac{K_I}{K_D}} \,. $$

A root locus sketch of a space station with a PID controller is shown in Figure 7.8. From Figure 7.8 we can see that as K_D varies from $0 < K_D < \infty$, the loci move from the origin into the left half-plane as desired. The selection of the ratios K_P/K_D and K_I/K_D will be influenced by the performance specifications.

If you think we concocted this space station discussion to make a point, a very similar analysis can be found in the report *Space Station Reference Configuration Description* [53]. This report details the preliminary design studies conducted early in the space station program at NASA. The original station was

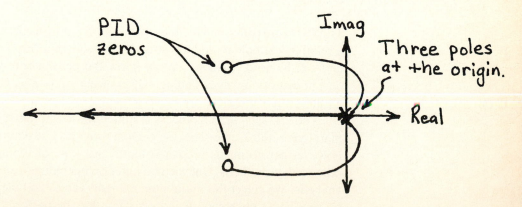

FIGURE 7.8
Root locus of the space station example with a PID controller.

dubbed the Power Tower. The values used in the preliminary analysis were

$$J = 7.81\text{E} + 07 \text{ slug} - \text{ft}^2 \, ,$$
$$K_P = 6.7\text{E} + 05 \, ,$$
$$K_D = 1.2\text{E} + 08 \, ,$$
$$K_I = 5.1\text{E} + 03 \, .$$

These values correspond to the pitch axis analysis; other values were used for the other axes—roll and yaw. These preliminary PID controller design studies were conducted by engineers in the so-called Skunk Works area at the NASA Johnson Space Center in the early 1980s. Those control engineers knew how to sketch a root locus by hand, and they performed much of the preliminary analysis with hand sketching and back-of-the-envelope discussion. Discussions of this type are extremely important in developing intuition and in making judgments regarding the applicability of various controller structures.

The advent of user-friendly computer systems and computer symbolic manipulation may someday relegate the argument for root locus hand sketching to the trash heap, along with the argument for the slide rule. It is the opinion of this author, however, that for the time being it is important to know how to hand-sketch a root locus. Take the time to learn the basic guidelines for sketching a root locus. A good discourse on the subject is in Chapter 7 of *Modern Control Systems*.

7.3 PID CONTROLLER

The **PID controller** is widely used in industry. Although the PID controller has a relatively simple three-term structure, as shown in Eq. (7.2), it is very robust and can be used when we do not know the plant model very accurately. The design process involves finding values of the gains K_P, K_D, and K_I such that all design specifications are satisfied.

The PID gain search process is sometimes referred to as **tuning**. For certain plant types, the Ziegler-Nichols method provides a structured tuning procedure; however not all plants are amenable to that type of tuning. Generally we use other ad hoc methods. One method relies on interactive computer simulation and analysis. We select the initial nominal gain values and then tune the controller by simulating the response and verifying that the performance is acceptable. If the performance is not acceptable, we update the gains and iterate the process. Naturally MATLAB is very helpful in this process.

7.3.1   Proportional Controller

If we set $K_D = K_I = 0$, the PID controller reduces to a **proportional controller**:

$$G_c(s) = K_P .$$

The proportional controller can be viewed as a volume control knob, which can be adjusted higher or lower. As discussed in Chapter 4 (see Section 4.2), raising the gain helps reduce the effects of disturbances and reduces the system sensitivity to plant parameter changes. However we cannot reject disturbances entirely, and steady-state errors will generally persist. Also raising the gain too much can result in closed-loop system instability and magnification of any measurement noise present in the system. On the plus side, the proportional controller is simple.

7.3.2   Integral Controller

If we set $K_P = K_D = 0$, then the PID controller reduces to an **integral controller**:

$$G_c(s) = \frac{K_I}{s} .$$

The benefits associated with the integral controller include reducing or eliminating steady-state tracking errors. This characteristic is due to the fact that using the integral controller increases the system type by one. So a type zero plant with an integral controller can track a step input with zero steady-state tracking error, a type one plant with an integral controller can track a ramp input with zero steady-state error, and so on.

We can put the proportional controller and integral controller together to form a PI controller:

$$G_c(s) = K_P + \frac{K_I}{s} = \frac{K_P s + K_I}{s} = K_P \left(\frac{s + \frac{K_I}{K_P}}{s} \right) .$$

The PI controller has one zero at $s = -K_I/K_P$ and a pole at the origin $s = 0$.

7.3.3   Derivative Controller

When $K_P = K_I = 0$, we have a **derivative controller**,

$$G_c(s) = K_D s .$$

We use a derivative controller to increase system damping and to improve stability.

Generally we use a derivative controller in conjunction with the proportional and/or integral terms. The PD controller is given by

$$G_c(s) = K_P + K_D s \,.$$

7.3.4   Cascade Controller: PID

When all three terms are included, the PID controller results (see Eq. 7.2). The PID controller reduces the steady-state error and improves the transient response. In fact, we can view the PID controller as a cascade of PI and PD controllers. Consider

$$G_{PI}(s) = \hat{K}_P + \frac{\hat{K}_I}{s}$$

and

$$G_{PD}(s) = \overline{K}_P + \overline{K}_D s \,,$$

where $\hat{K}_I$ and $\hat{K}_P$ are the PI controller gains and $\overline{K}_P$ and $\overline{K}_D$ are the PD controller gains. Then cascading the two controllers yields

$$G_c(s) = G_{PI}(s)G_{PD}(s)$$

$$= \left(\hat{K}_P + \frac{\hat{K}_I}{s} \right) \left(\overline{K}_P + \overline{K}_D s \right)$$

$$= \left(\hat{K}_I \overline{K}_D + \hat{K}_P \overline{K}_P \right) + \hat{K}_P \overline{K}_D s + \frac{\hat{K}_I \overline{K}_P}{s}$$

$$= K_P + K_D s + \frac{K_I}{s} \,,$$

where we have the relationships

$$K_P = \hat{K}_I \overline{K}_D + \hat{K}_P \overline{K}_P \,,$$
$$K_D = \hat{K}_P \overline{K}_D \,,$$
$$K_I = \hat{K}_I \overline{K}_P \,.$$

7.4   AUTOMOBILE VELOCITY CONTROL

Electronic systems currently make up about 6% of a car's value. That figure will climb to 20% by the year 2000 as antilock brakes, active suspensions, and other computer-dependent technologies move into full production. Much of the added computing power will be used for new technology for smart cars and smart roads, such as IVHS (intelligent vehicle/highway systems) [54], [55].

The term IVHS refers to a varied assortment of electronics that provides real-time information on accidents, congestion, and roadside services to drivers and traffic controllers. IVHS also encompasses devices that make vehicles more autonomous: collision-avoidance systems and lane-tracking technology that alert drivers to impending disasters and allow a car to drive itself.

An example of an automated highway system is shown in Figure 7.9. This example is adapted from DP7.12 in *Modern Control Systems*. A velocity control system for maintaining the velocity between vehicles is shown in Figure 7.10. The output, $Y(s)$, is the relative velocity of the two automobiles; the input, $R(s)$, is the desired relative velocity between the two vehicles. Our design goal is to develop a controller that can maintain the prescribed velocity between the vehicles and maneuver the active vehicle (in this case the rearward automobile) as commanded.

The control goal is

Control Goal   Maintain the prescribed velocity between the two vehicles, and maneuver the active vehicle as commanded.

The variable to be controlled is the relative velocity between the two vehicles:

Variable to Be Controlled   The relative velocity between vehicles, denoted by $y(t)$.

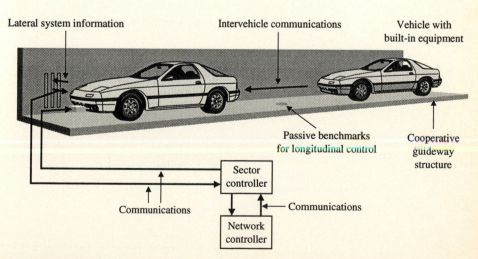

FIGURE 7.9
Automated highway system.

The design specifications are

DS1 Zero steady-state error to a step input.

DS2 Steady-state error due to a ramp input of less than 25% of the input magnitude.

DS3 Percent overshoot less than 5% to a step input.

DS4 Settling time less than 1.5 seconds to a step input (using a 2% criterion to establish settling time).

From the design specifications and knowledge of the open-loop system, we find that we need a **type one system** to guarantee a zero steady-state error to a step input. The open-loop system transfer function is a type zero system; therefore, the controller needs to increase the system type by at least 1. A type one controller (that is, a controller with one integrator) satisfies DS1. To meet DS2 we need to have

$$K_v = \lim_{s \to 0} s G_c(s) G(s) \geq \frac{1}{0.25} = 4 \,, \tag{7.3}$$

where

$$G(s) = \frac{1}{(s + 2)(s + 8)} \,, \tag{7.4}$$

and $G_c(s)$ is the controller (yet to be specified).

The percent overshoot specification, DS3, allows us to define a target damping ratio:

$$P.O. \leq 5\% \quad \text{implies} \quad \zeta \geq 0.69 \,.$$

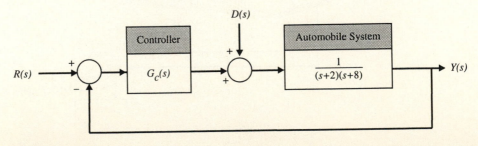

FIGURE 7.10
Vehicle velocity control system.

Similarly from the settling time specification, DS4, we have

$$T_s \approx \frac{4}{\zeta \omega_n} \leq 1.5 .$$

Solving for $\zeta \omega_n$ yields

$$\zeta \omega_n \geq 2.66 .$$

The desired region for the poles of the closed-loop transfer function is shown in Figure 7.11. Using a proportional controller

$$G_c(s) = K_P ,$$

is not reasonable, because DS2 cannot be satisfied. We need at least one pole at the origin to track a ramp input. Consider the PI controller

$$G_c(s) = \frac{K_P s + K_I}{s} = K_P \left(\frac{s + \frac{K_I}{K_P}}{s} \right) . \tag{7.5}$$

The question is where to place the zero at $s = -K_I / K_P$.

   For what values of K_P and K_I is the system stable? The closed-loop transfer function is

$$T(s) = \frac{K_P s + K_I}{s^3 + 10s^2 + (16 + K_P)s + K_I} .$$

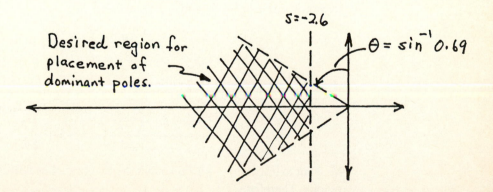

FIGURE 7.11
Desired region in the complex plane for locating the dominant system poles.

The corresponding Routh array is

$$
\begin{array}{c|cc}
s^3 & 1 & 16+K_P \\
s^2 & 10 & K_I \\
s & \frac{10(K_P+16)-K_I}{10} & 0 \\
1 & K_I &
\end{array}
$$

The first requirement for stability (from column one, row four) is

$$K_I > 0. \tag{7.6}$$

From the first column, third row, we have the inequality

$$K_P > \frac{K_I}{10} - 16. \tag{7.7}$$

It follows from DS2 that

$$K_v = \lim_{s \to 0} sG_c(s)G(s) = \lim_{s \to 0} s\frac{K_P\left(s + \frac{K_I}{K_P}\right)}{s} \frac{1}{(s+2)(s+8)} = \frac{K_I}{16} > 4.$$

Therefore, the integral gain must satisfy

$$K_I > 64. \tag{7.8}$$

If we select $K_I > 64$, then the inequality in Eq. (7.6) is satisfied. The valid region for K_P is then given by Eq. 7.7, where $K_I > 64$.

We need to consider DS4. Here we want to have the dominant poles to the left of the $s = -2.66$ line. We know from our experience sketching the root locus that since we have three poles (at $s = 0, -2,$ and -8) and one zero (at $s = -K_I/K_P$), we expect two branches of the loci to go to infinity along two asymptotes at $\phi = -90°$ and $+90°$ centered at

$$\alpha = \frac{\sum p_i - \sum z_i}{n_p - n_z},$$

where $n_p = 3$ and $n_z = 1$. In our case

$$\alpha = \frac{-2 - 8 - \left(-\frac{K_I}{K_P}\right)}{2} = -5 + \frac{1}{2}\frac{K_I}{K_P}.$$

We want to have $\alpha < -2.66$ so that the two branches will bend into the desired regions. Therefore,

$$-5 + \frac{1}{2}\frac{K_I}{K_P} < -2.66,$$

or

$$\frac{K_I}{K_P} < \frac{14}{3} \, . \qquad (7.9)$$

So as a first design, we can select K_P and K_I such that

$$K_I > 64 \, ,$$

$$K_P > \frac{K_I}{10} - 16 \, ,$$

$$\frac{K_I}{K_P} < \frac{14}{3} \, .$$

Suppose we choose

$$\frac{K_I}{K_P} = 2.5 \, .$$

Then the closed-loop characteristic equation is

$$1 + K_P \frac{(s + 2.5)}{s(s + 2)(s + 8)} = 0 \, .$$

The root locus is shown in Figure 7.12. To meet the $\zeta = 0.69$ (which evolved from DS3), we need to select

$$K_P < 30 \, .$$

We obtain this value from the root locus using the rlocfind command. We selected the value at the boundary of the performance region (see Figure 7.12) as carefully as possible. Nevertheless, the displayed value of $K_P = 30.1079$ is approximate. The value K_P above is rounded-off from the value displayed by MATLAB.

Selecting $K_P = 26$, we have

$$\frac{K_I}{K_P} = 2.5 \quad \Rightarrow \quad K_I = 65 \, .$$

This satisfies the steady-state tracking error specification (DS2) since $K_I = 65 > 64$.

The resulting PI controller is

$$G_c(s) = 26 + \frac{65}{s} \, . \qquad (7.10)$$

The step response is shown in Figure 7.13.

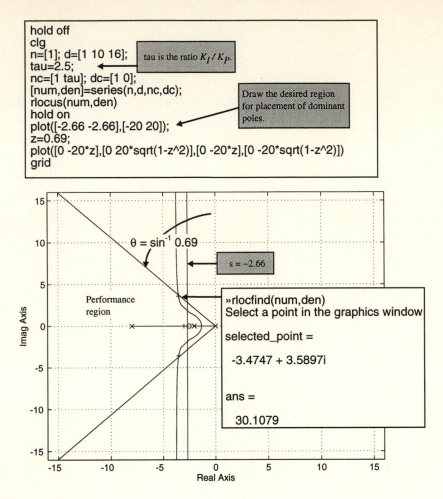

```
hold off
clg
n=[1]; d=[1 10 16];
tau=2.5;                          tau is the ratio $K_I/K_P$.
nc=[1 tau]; dc=[1 0];
[num,den]=series(n,d,nc,dc);       Draw the desired region
rlocus(num,den)                    for placement of dominant
hold on                            poles.
plot([-2.66 -2.66],[-20 20]);
z=0.69;
plot([0 -20*z],[0 20*sqrt(1-z^2)],[0 -20*z],[0 -20*sqrt(1-z^2)])
grid
```

$\theta = \sin^{-1} 0.69$

$s = -2.66$

Performance region

```
»rlocfind(num,den)
Select a point in the graphics window

selected_point =

   -3.4747 + 3.5897i

ans =

   30.1079
```

FIGURE 7.12
Root locus for $K_I/K_P = 2.5$.

The percent overshoot is

$$P.O. = 8\% \,,$$

and the settling time is

$$T_s = 1.45 \text{ sec} \,.$$

The percent overshoot specification is not precisely satisfied, but the controller in Eq. (7.10) represents a very good first design. We can iteratively refine it using MATLAB. Even though the closed-loop poles lie in the desired region, the response does not exactly meet the specifications because the compensator zero influences the response. The closed-loop system is a third-order system and does not have the performance of a second-order system. We might consider moving

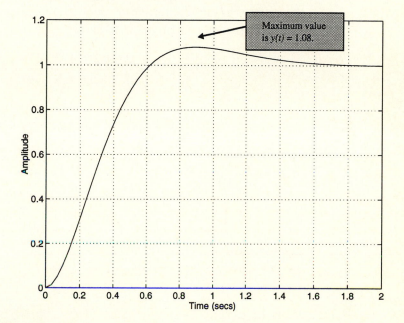

```
hold off, clg
n=[1]; d=[1 10 16];
Kp=26;Ki=65;
nc=[Kp Ki]; dc=[1 0];
[numa,dena]=series(n,d,nc,dc);
[na,da]=cloop(numa,dena);
step(na,da), grid
```

FIGURE 7.13
Automobile velocity control using the PI controller in Eq. (7.10).

the zero to $s = -2$ (by choosing $K_I/K_P = 2$) so that the pole at $s = -2$ is cancelled and the resulting system is truly a second-order system.

7.5   SUMMARY

In this chapter we designed a PI controller to meet the specifications associated with the automobile velocity control system. By sketching the root locus we determined that the controller needed to possess a zero in the left half-plane to attract the roots into the desired region to meet the design specifications. We used the root locus design method to obtain valid values of the controller gains. The designed controller nearly met all the specifications on the first iteration, but further refinement is necessary to lower the percent overshoot slightly.

EXERCISES

E7.1   Using the parameter values for the Power Tower given in Section 7.2, determine the settling time for the spacecraft to maneuver from a $10°$ pitch attitude to $0°$ attitude.

E7.2   Design a controller to meet the design specifications given in Section 7.4 when the system has the additional time-delay shown in Figure E7.2. The time-delay is $T = 0.25$ seconds. The time-delay can be approximated by the transfer function

$$G_{TD}(s) = \frac{0.0052s^2 - 0.1240s + 0.9923}{0.0052s^2 + 0.1240s + 0.9923}.$$

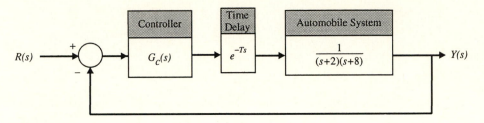

FIGURE E7.2
Vehicle velocity control system with a time-delay.

E7.3   What happens to the performance of the velocity control system designed in Section 7.4 if the parameters of the automobile change? The transfer function is given by

$$G(s) = \frac{1}{(s + a)(s + b)}.$$

The nominal parameter values are $a = 2$ and $b = 8$. Consider the PI controller

$$G_c(s) = K_P + \frac{K_I}{s} = 33 + \frac{66}{s}. \tag{7.11}$$

Verify that the PI controller in Eq. (7.11) meets the four design specifications given in Section 7.4 with $a = 2$ and $b = 8$. With the PI controller in Eq. (7.11), how much can a and b vary yet allow the system to remain stable and meet the design specifications? Are there other values for K_P and K_I in Eq. (7.11) such that the closed-loop system is more robust to changes in the plant parameters? By robust we mean the system remains stable and meets the performance specifications for a wider range of a and b.

Frequency Response Methods

Six-legged Ambler Example

8.1 Introduction   140

8.2 A Simple Physical System   142

8.3 Six-legged Ambler   146

8.4 Controller Selection   147

8.5 Controller Design   148

8.6 Summary   154

Exercises   155

P R E V I E W

The frequency response of linear time-invariant systems is the main topic of this chapter. We begin by considering a simple mass-spring-damper system and show that the steady-state response of the system output to a sinusoidal input is also sinusoidal. The frequency of the ouput is the same as the frequency of the input signal. However, the magnitude and phase of the output signal differs from the input signal, and the amount of difference is a function of the input frequency. In fact, a plot of the magnitude versus input frequency and phase shift versus input frequency is known as a Bode plot. We use the Ambler mobile robot to illustrate the use of MATLAB in frequency response analysis. We design a controller to meet both bandwidth and percent overshoot specifications.

8.1   INTRODUCTION

What happens when we drive a physical system with a sinusoidal input? Our intuition might tell us that the output would also be sinusoidal. But is there a relationship between the magnitude of the sinusoidal input and the magnitude of the output? A related question is, What is the phase difference between the input signal and the resulting output? We might also wonder how the physical system responds as the sinusoidal input frequency gets higher and higher. Can a physical system respond to a very high frequency input? Finally, we might like to know how the response of a system to a sinusoidal input signal relates to the issue of control system design. These topics are the focus of this chapter. The elements of the design process emphasized in the chapter are shown in Figure 8.1.

A **Bode plot** is a graphical representation of the magnitude versus frequency and phase shift versus frequency. We can obtain a Bode plot by hand-sketching or by using MATLAB. If MATLAB does a good job producing Bode plots (and it does), why worry about learning how to sketch the plot by hand? The same arguments for sketching root loci (see the discussions in Chapter 7 on root locus sketching) apply for sketching Bode plots. Knowing how to sketch a Bode plot gives us an additional capability to help in the process of selecting a controller to shape the Bode plot in a desired fashion. In the end, it is always about design! Many practicing engineers express their dismay at the lack of ability of new engineering graduates to perform simple back of the envelope design and analyses. Without the latest high-performance computing machines, these new control engineers are lost. And even when the computing facilities are available, the ability to perform satisfactory **reasonableness checks** is absent.[1]

Consider the following situation. A company has a test rig in which a heavy object is supported by six actuators that are of the ball screw variety. We are tasked to design, implement, and test an actuator control system maneuvering a heavy object in a precise manner. The manufacturer has provided a brief written description of the actuators (no mathematical models are provided). A one-page technical description of the driving motors that provide the input to the actuators is also available. No information is provided on the position sensors, except that the measurements will be noisy. The control bandwidth is specified to be 10 Hz. A deadline of three weeks is set to finish the task—the final qualification of the test article cannot proceed without the control system!

Is this an unrealistic scenario? No! This scenario is based on an actual event. One approach to the design problem is to use an idealized motor transfer function and a simple second-order model of the actuator. Glean whatever parameter values (motor constants, mass of moving actuator parts, equivalent spring

---

1. Based on the author's personal discussions with Boeing Company engineers during the summer of 1996.

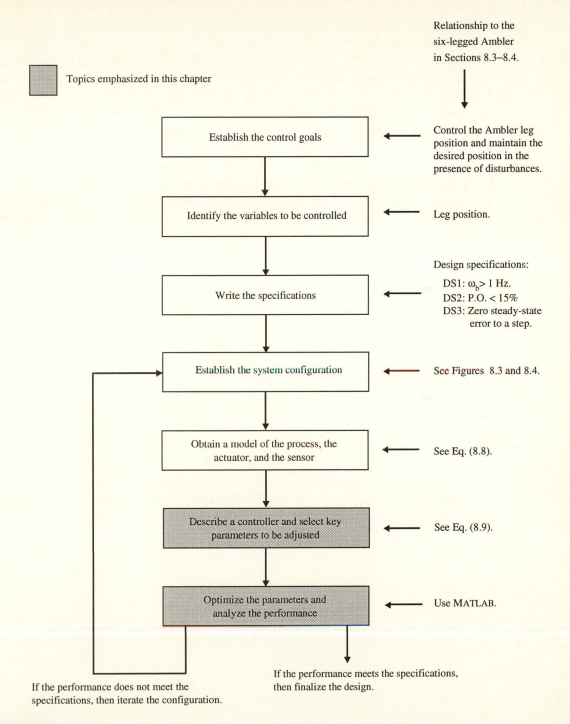

FIGURE 8.1
Elements of the control system design process emphasized in this chapter.

provided. We may have to go into the laboratory to test the open-loop system to determine any unknown parameters, or we may have to provide a test matrix to the engineers in the laboratory who will then perform the tests (written and oral communication are important). Then sketch a Bode plot of the system and determine how to modify the shape of the Bode plot to meet the 10 Hz bandwidth specification. The ability to sketch the Bode plot lets us determine quickly a number of candidate controllers and important control parameters. So far we have not needed a computer, but a MATLAB/SIMULINK simulation would be a logical next step. We develop a simulation of the closed-loop system and determine useful values of the various controller parameters. Using the latest in rapid-prototyping software/hardware, we implement the candidate controllers in the laboratory, test, and fine-tune. The laboratory experience will generally be useful in increasing the fidelity and accuracy of the SIMULINK simulation (we may need to add deadbands, rate limits, more accurate actuator models, and so on).

There is a critical synergy between back-of-the-envelope engineering and high-performance computing. If you cannot readily sketch Bode plots by hand, take the time to practice. It may be that the idea of hand-sketching will someday become history. As computer programs get more and more user-friendly it may make more sense to sketch with the assistance of a computer. That day has yet to arrive, so we suggest you develop the ability to perform back-of-the-envelope computations using hand-sketched Bode plots.

8.2   A SIMPLE PHYSICAL SYSTEM

It is sometimes very useful to consider a simple physical system when initially developing a feel for a problem. To this end we consider the simple mass-spring-damper system shown in Figure 8.2. Summing the forces acting on the mass m,

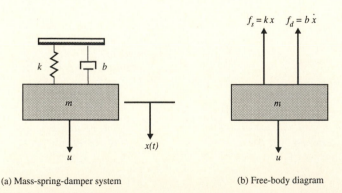

(a) Mass-spring-damper system   (b) Free-body diagram

FIGURE 8.2
Simple mass-spring-damper system.

we have

$$\sum f = u + f_s + f_d = u - kx - b\dot{x} \, ,$$

where u is the input driving force. Then using Newton's second law ($\sum f = m\ddot{x}$), it follows that the equation governing the motion of the mass is

$$m\ddot{x} + b\dot{x} + kx = u \, . \qquad (8.1)$$

Suppose $u(t)$ is a sinusoidal input:

$$u(t) = a \sin \omega t \, . \qquad (8.2)$$

The magnitude of the input is a, and the frequency is ω. The transfer function of the system in Eq. (8.1) is

$$G(s) = \frac{X(s)}{U(s)} = \frac{1/m}{s^2 + \bar{b}s + \bar{k}} \, , \qquad (8.3)$$

where

$$\bar{b} = \frac{b}{m} \, ,$$

$$\bar{k} = \frac{k}{m} \, .$$

Assume that the mass-spring-damper system is such that

$$\bar{k} - \bar{b}^2/4 > 0 \, .$$

This corresponds to an effective system damping ratio $\zeta < 1$.

The Laplace transform of the sinusoidal input (see Appendix A in *Modern Control Systems*) is

$$U(s) = \frac{a\omega}{s^2 + \omega^2} \, .$$

Thus,

$$X(s) = \frac{a\omega/m}{(s^2 + \bar{b}s + \bar{k})(s^2 + \omega^2)} \, . \qquad (8.4)$$

Expanding Eq. (8.4) in a **partial fraction expansion** yields

$$X(s) = \frac{c_1 s + c_o}{s^2 + \bar{b}s + \bar{k}} + \frac{d_1 s + d_o}{s^2 + \omega^2} \qquad (8.5)$$

where

$$c_o = \frac{a\omega(\omega^2 - \bar{k} + \bar{b}^2)}{m\Delta^2} \, ,$$

$$c_1 = \frac{a\overline{b}\omega}{m\Delta^2} \, ,$$

$$d_o = \frac{a\omega(\overline{k} - \omega^2)}{m\Delta^2} \, ,$$

$$d_1 = -\frac{a\overline{b}\omega}{m\Delta^2} \, ,$$

$$\Delta = \sqrt{(\overline{k} - \omega^2)^2 + \overline{b}^2\omega^2} \, .$$

Taking the **inverse Laplace transform** of Eq. (8.5) yields the time-domain solution, $x(t)$, due to a sinusoidal input:

$$x(t) = \frac{a}{m\Delta}\left[sin(\omega t + \phi) + \Gamma e^{-\frac{\overline{b}}{2}t}sin(\Omega t + \theta)\right] , \qquad (8.6)$$

where

$$\phi = \tan^{-1}\frac{\omega\overline{b}}{\omega^2 - \overline{k}} \, ,$$

$$\Omega = \sqrt{\overline{k} - \frac{\overline{b}^2}{4}} \, ,$$

$$\theta = \tan^{-1}\frac{\overline{b}\sqrt{\overline{k} - \overline{b}^2/4}}{\omega^2 - \overline{k} + \overline{b}^2/2} \, ,$$

$$\Gamma = \frac{\overline{b}\omega}{\Delta}\left[\frac{1}{\overline{k} - \frac{1}{4}\overline{b}^2}\left[\frac{\omega^2 - \overline{k} + \frac{1}{2}\overline{b}^2}{\overline{b}}\right]^2 + 1\right]^{\frac{1}{2}} .$$

We can obtain several interesting insights by examining Eq. (8.6). For nonzero damping, $\overline{b} > 0$, we have

$$x(t) \;\rightarrow\; \frac{a}{m\Delta}\sin(\omega t + \phi) \quad \text{as} \quad t \rightarrow \infty \, .$$

In the steady-state, the time-history of the position of the mass is sinusoidal with magnitude $a/(m\Delta)$ (recall that the input magnitude is a). Additionally, the frequency of the time-history position of the mass is ω, the same as the input sinusoidal frequency. However, the position time-history is out of phase with the input signal by the angle

$$\phi = \tan^{-1}\frac{\omega\overline{b}}{\omega^2 - \overline{k}} \, . \qquad (8.7)$$

The amount of attenuation and/or magnification of the input

$$|G(j\omega)| = \frac{1}{m\Delta} = \frac{1}{m\sqrt{(\bar{k} - \omega^2)^2 + \bar{b}^2\omega^2}} \, ,$$

and the amount of phase shift [given by ϕ in Eq. (8.7)] are dependent on the input frequency ω. If we plot $|G(j\omega)|$ versus ω and ϕ versus ω, we have a Bode plot.

What happens to the position time-history (in the steady-state) as the input frequency, ω, increases? We have

$$\Delta = \sqrt{(\bar{k} - \omega^2)^2 + \bar{b}^2\omega^2} \quad \Rightarrow \quad \Delta \approx \omega^2 \quad \text{as} \quad \omega \to \infty \, .$$

The position output magnitude (which is inversely proportional to Δ) goes to zero as $\omega \to \infty$. Therefore, we say that the mass-spring-damper system does not respond significantly to input signals of very high frequencies. This is true in general for other physical systems as well.

Suppose the mass-spring-damper system has very little damping:

$$\bar{b} << 1 \, .$$

Large space structures generally exhibit minimal damping. Then in the steady-state we have

$$x(t) \approx \frac{a}{m(\bar{k} - \omega^2)} \sin \omega t \, .$$

If we select a sinusoidal input with frequency near $\omega = \sqrt{k/m}$, the output magnitude gets large (that is, the system resonates). The result is usually some type of system failure (for example, consider the Tacoma Narrows Bridge, described in Chapter 6 of *Modern Control Systems*, for photographic evidence). The characteristics of the mass-spring-damper system response to a sinusoidal input applies to many other linear systems [56].

The **frequency response** of a stable linear system is defined to be the steady-state response of the output to a sinusoidal input. In the steady-state, the output signal differs from the input signal only in amplitude and phase angle—the frequency remains unchanged. The attenuation and/or amplification of the input sinusoidal signal and the phase angle are both functions of the input frequency. As we have just seen for the mass-spring-damper system, as the frequency of the input sinusoidal signal increases to large values, the output signal is significantly attenuated—physical systems generally cannot respond to very high frequency inputs. The amount of attenuation depends on the individual system.

8.3   SIX-LEGGED AMBLER

The Ambler is a six-legged walking machine being developed at Carnegie-Mellon University [57]. An artist's conception of the Ambler is shown in Figure 8.3. This problem is adapted from DP8.2 in *Modern Control Systems*.

In this section we consider the control system design for position control of one leg. The mathematical model of the actuator and leg is provided. The transfer function is

$$G(s) = \frac{1}{s(s^2 + 2s + 10)} \, . \tag{8.8}$$

The input is a voltage command to the actuator, and the output is the leg position (vertical position only). A block diagram of the control system is shown in Figure 8.4. The control goal is

---

Control Goal   Control the Ambler leg position and maintain the position in the presence of unwanted measurement noise.

The variable to be controlled is

Variable to Be   Leg position, $Y(s)$.
Controlled

We want the leg to move to the commanded position as fast as possible but with minimal overshoot. As a practical first step, the design goal will be to produce a system that moves, albeit slowly. In other words, the control system bandwidth

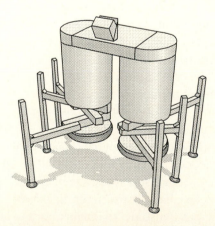

FIGURE 8.3
An artist's conception of the six-legged Ambler.

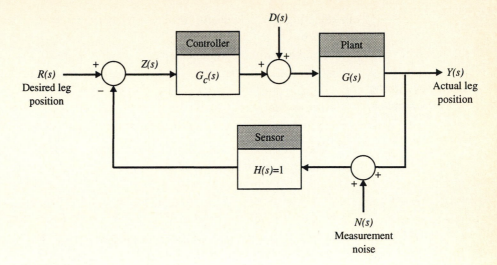

FIGURE 8.4
Ambler control system for one leg.

will initially be low—around 1 Hz. We consider the problem of increasing the bandwidth to be a future project.

The control design specifications are

Control Design
Specifications

DS1 Closed-loop bandwidth greater than 1 Hz.

DS2 Percent overshoot less than 15% to a step input.

DS3 Zero steady-state tracking error to a step input.

Specifications DS1 and DS2 are intended to ensure acceptable tracking performance. Design specification DS3 is actually a nonissue in our design: the actuator/leg transfer function is a type one system so a zero steady-state tracking error to a step input is guaranteed. We simply need to ensure that $G_c(s)G(s)$ remains at least a type one system.

8.4 CONTROLLER SELECTION

Consider the PID controller

$$G_c(s) = \frac{K(s^2 + as + b)}{s + c} .$$

(8.9)

As $c \to 0$, a PID controller is obtained with $K_P = Ka$, $K_D = K$, and $K_I = Kb$. We can let c be a parameter at this point and see if the additional freedom in selecting $c \neq 0$ is useful. It may be that we can simply set $c = 0$ and use the PID form.

The key tuning parameters are

Select Key Tuning Parameters

K, a, b and c.

The controller in Eq. (8.9) is not the only controller that we can consider. For example, we might consider

$$G_c(s) = K \frac{s + z}{s + p} \, , \tag{8.10}$$

where K, z, and p are the key tuning parameters. The design of the type of controller, given in Eq. (8.10) will be left as an exercise at the end of the chapter.

8.5   CONTROLLER DESIGN

The response of a closed-loop control system is determined predominantly by the location of the **dominant poles**. Our approach to the design is to determine appropriate locations for the dominant poles of the closed-loop system. We determine the locations from the performance specifications by using second-order system approximation formulas. We obtain the controller parameters so that the closed-loop system has the desired dominant poles, and the remaining poles are located such that their contribution to the overall response is negligible.

The bandwidth, ω_B, is approximately related to the natural frequency, ω_n, by

$$\frac{\omega_B}{\omega_n} \approx -1.1961\zeta + 1.8508 \quad (0.3 \leq \zeta \leq 0.8) \, . \tag{8.11}$$

This approximation applies to second-order systems. For more discussion on this and other second-order system relationships, see Chapter 8 in *Modern Control Systems*. Per specification DS1, we want

$$\omega_B = 1 \text{ Hz} = 6.28 \text{ rad/sec} \, . \tag{8.12}$$

From the overshoot specification, we can determine the minimum value of ζ. Thus for $P.O. \leq 15\%$, we require

$$\zeta \geq 0.52 \, , \tag{8.13}$$

where we have used the fact (valid for second-order systems) that

$$P.O. = 100e^{\frac{-\pi\zeta}{\sqrt{1-\zeta^2}}} \, .$$

Another useful design formula relates $M_{p_\omega} = |T(\omega_r)|$ to the damping ratio:

$$M_{p_\omega} = |T(\omega_r)| = \frac{1}{2\zeta\sqrt{1-\zeta^2}} \quad (\zeta < 0.707) . \tag{8.14}$$

The relationship between the resonant frequency, ω_r, the natural frequency, ω_n, and the damping ratio, ζ, is given by

$$\omega_r = \omega_n\sqrt{1-\zeta^2} \quad (\zeta < 0.707) . \tag{8.15}$$

We require $\zeta \geq 0.52$; therefore, we will design with $\zeta = 0.52$. We select ζ on the boundary because, in general, selecting larger ζ results in a larger settling time. Even though settling time is not a design specification for this problem, we usually attempt to make the system response as fast as possible while still meeting all the design specifications. From Eqs. (8.11) and (8.12) it follows that

$$\omega_n = \frac{\omega_B}{-1.1961\zeta + 1.8508} = 5.11 \text{ rad/sec} . \tag{8.16}$$

The expected 2% settling time is approximately

$$T_s \approx \frac{4}{\zeta\omega_n} = 1.5 \text{ sec} .$$

Then with $\omega_n = 5.11$ rad/sec and $\zeta = 0.52$ and using Eq. (8.15) we compute

$$\omega_r = 4.36 \text{ rad/sec} . \tag{8.17}$$

So if we had a second-order system, we would want to determine values of the control gains such that

$$\omega_n = 5.11 \text{ rad/sec} \quad \text{and} \quad \zeta = 0.52 ,$$

or equivalently,

$$M_{p_\omega} = 1.125 \quad \text{and} \quad \omega_r = 4.36 \text{ rad/sec} .$$

Our closed-loop system is a fourth-order system and not a second-order system. So, a valid design approach would be to select K, a, b, and c so that two poles are dominant and located appropriately to meet the design specifications. This will be the approach followed here.

Another valid approach is to develop a second-order approximation of the fourth-order system. In the approximate transfer function, the parameters K, a, b, and c are left as variables. Following the approach discussed in Chapter 5 of *Modern Control Systems*, we can obtain the approximate transfer function, $T_a(s)$, in such a way that the frequency response of $T_a(s)$ is very close to that of the original system $T(s)$.

The open-loop transfer function is

$$G_c(s)G(s) = \frac{K(s^2 + as + b)}{s(s^2 + 2s + 10)(s + c)},$$

and the closed-loop transfer function is

$$T(s) = \frac{G_c(s)G(s)}{1 + G_c(s)G(s)} \tag{8.18}$$

$$= \frac{K(s^2 + as + b)}{s^4 + (2 + c)s^3 + (10 + 2c + K)s^2 + (10c + Ka)s + Kb}.$$

The associated characteristic equation is

$$s^4 + (2 + c)s^3 + (10 + 2c + K)s^2 + (10c + Ka)s + Kb = 0. \tag{8.19}$$

The desired characteristic polynomial must also be fourth-order, but we want it to be composed of multiple factors, as follows:

$$p_d(s) = (s^2 + 2\zeta\omega_n s + \omega_n^2)(s^2 + d_1 s + d_o),$$

where ζ and ω_n are selected to meet the design specifications, and the roots of $s^2 + 2\zeta\omega_n s + \omega_n^2 = 0$ are the dominant roots. Conversely we want the roots of $s^2 + d_1 s + d_o = 0$ to be the nondominant roots. The dominant roots should lie on a vertical line in the complex plane defined by the distance $s = -\zeta\omega_n$ away from the imaginary axis. Let

$$d_1 = 2\alpha\zeta\omega_n.$$

Then the roots of $s^2 + d_1 s + d_o = 0$ lie on a vertical line in the complex plane defined by $s = -\alpha\zeta\omega_n$. By choosing $\alpha > 1$, we effectively move the roots to the left of the dominant roots. The larger we select α, the further the nondominant roots lie to the left of the dominant roots. A reasonable value of α is

$$\alpha = 10.$$

We can easily experiment with different values using MATLAB. Also, if we select

$$d_o = \alpha^2 \zeta^2 \omega_n^2,$$

then we obtain two real roots

$$s^2 + d_1 s + d_o = (s + \alpha\zeta\omega_n)^2 = 0.$$

Choosing $d_o = \alpha^2 \zeta^2 \omega_n^2$ is not required, but this seems to be a reasonable choice since we would like the contribution of the nondominant roots to the overall response to be quickly fading and nonoscillatory.

The desired characteristic polynomial is then

$$s^4 + [2\zeta\omega_n(1+\alpha)]s^3 + [\omega_n^2(1+\alpha\zeta^2(\alpha+4))]s^2 \qquad (8.20)$$
$$+ [2\alpha\zeta\omega_n^3(1+\zeta^2\alpha)]s + \alpha^2\zeta^2\omega_n^4 = 0 \, .$$

Equating the coefficients of Eqs. (8.19) and (8.20) yields four relationships involving $K, a, b, c,$ and α:

$$2\zeta\omega_n(1+\alpha) = 2+c \, ,$$
$$\omega_n^2(1+\alpha\zeta^2(4+\alpha)) = 10 + 2c + K \, ,$$
$$2\alpha\zeta\omega_n^3(1+\zeta^2\alpha) = 10c + Ka \, ,$$
$$\alpha^2\zeta^2\omega_n^4 = Kb \, .$$

In our case $\zeta = 0.52$, $\omega_n = 5.11$, and $\alpha = 10$. Thus we obtain

$$c = 56.46 \, ,$$
$$K = 891.69 \, ,$$
$$a = 5.14 \, ,$$
$$b = 20.68 \, ,$$

and the resulting controller is

$$G_c(s) = 891.69\frac{s^2 + 5.14s + 20.68}{s + 56.46} \, . \qquad (8.21)$$

The step response of the closed-loop system using the controller in Eq. (8.21) is shown in Figure 8.5. The percent overshoot is $P.O. = 16\%$, and the settling time is $T_s = 1$ second. These are relatively close to the desired values, but some fine-tuning of the controller will be necessary to reduce the $P.O.$ to less than 15%. The settling time is also a bit faster than expected. As a first iteration on the design, we can raise $\zeta = 0.52$ a little higher (off the boundary). This would lower the $P.O.$ and lengthen the settling time.

The magnitude plot of the closed-loop system is shown in Figure 8.6. The bandwidth is $\omega_b = 24.18$ rad/sec = 4.2 Hz. This satisfies DS1 but is larger than the $\omega_b = 1$ Hz used in the design (due to the fact that our system is not a second-order system). The higher bandwidth leads us to expect a faster settling time. In fact, the settling time is $T_s = 1$ second and we were expecting $T_s = 1.5$ seconds.

The peak magnitude is $M_{p\omega} = 1.24$. We were expecting $M_{p\omega} = 1.125$. The slightly higher peak magnitude leads us to expect a higher percent overshoot. This is the case with the actual $P.O. = 16\%$ and the expected $P.O. = 15\%$.

What is the steady-state response of the closed-loop system if the input is a sinusoidal input? From our previous discussions we expect that as the input frequency increases, the magnitude of the output will decrease. Two cases are presented here. In Figure 8.7 the input frequency is $\omega = 1$ rad/sec. The output

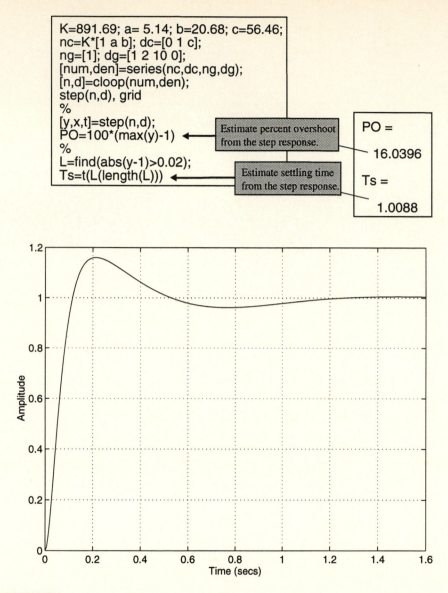

```
K=891.69; a= 5.14; b=20.68; c=56.46;
nc=K*[1 a b]; dc=[0 1 c];
ng=[1]; dg=[1 2 10 0];
[num,den]=series(nc,dc,ng,dg);
[n,d]=cloop(num,den);
step(n,d), grid
%
[y,x,t]=step(n,d);
PO=100*(max(y)-1)
%
L=find(abs(y-1)>0.02);
Ts=t(L(length(L)))
```

Estimate percent overshoot from the step response.

Estimate settling time from the step response.

PO =

16.0396

Ts =

1.0088

FIGURE 8.5
Step response using the controller in Eq. (8.21).

magnitude is approximately equal to 1 in the steady-state. In Figure 8.8 the input frequency is $\omega = 500$ rad/sec. The output magnitude is less than 0.005 in the steady-state. This verifies our intuition that the Ambler system (which is a physical system) responds less and less as the input sinusoidal frequency increases.

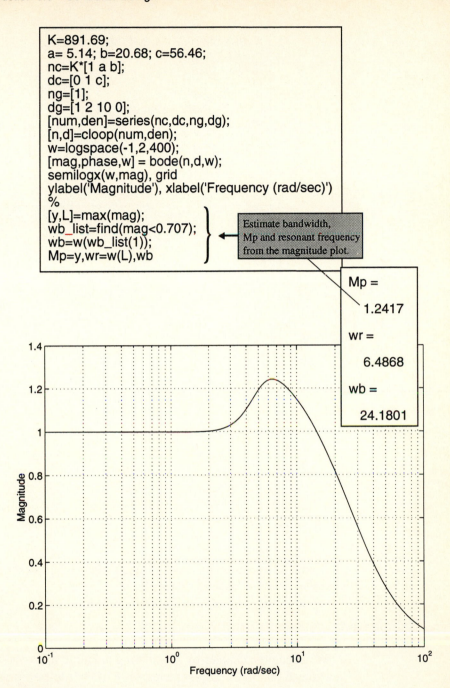

```
K=891.69;
a= 5.14; b=20.68; c=56.46;
nc=K*[1 a b];
dc=[0 1 c];
ng=[1];
dg=[1 2 10 0];
[num,den]=series(nc,dc,ng,dg);
[n,d]=cloop(num,den);
w=logspace(-1,2,400);
[mag,phase,w] = bode(n,d,w);
semilogx(w,mag), grid
ylabel('Magnitude'), xlabel('Frequency (rad/sec)')
%
[y,L]=max(mag);
wb_list=find(mag<0.707);
wb=w(wb_list(1));
Mp=y,wr=w(L),wb
```

Estimate bandwidth, Mp and resonant frequency from the magnitude plot.

Mp =

1.2417

wr =

6.4868

wb =

24.1801

FIGURE 8.6
Magnitude plot of the closed-loop system with the controller in Eq. (8.21).

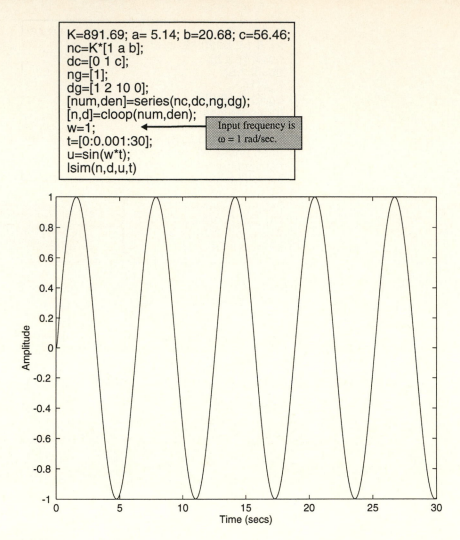

FIGURE 8.7
Output response of the Ambler closed-loop system when the input is a sinusoidal signal of frequency $\omega = 1$ rad/sec.

8.6 SUMMARY

We began the chapter by considering a simple mass-spring-damper system. We showed that the steady-state response of the mass-spring-damper system output to a sinusoidal input is also sinusoidal. We discussed the fact that the frequency of the ouput signal is the same as the frequency of the input signal but the magnitude and phase differs from the input signal. The amount of difference is a function of the input frequency. We were reminded that a plot of the magnitude versus input frequency and phase shift versus input frequency is known as a Bode plot. We designed a controller to meet both bandwidth and percent over-

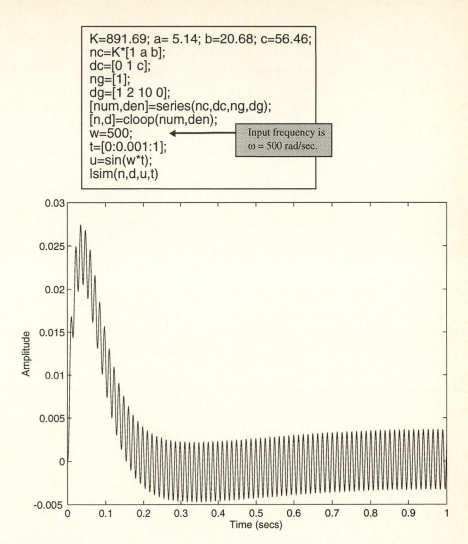

```
K=891.69; a= 5.14; b=20.68; c=56.46;
nc=K*[1 a b];
dc=[0 1 c];
ng=[1];
dg=[1 2 10 0];
[num,den]=series(nc,dc,ng,dg);
[n,d]=cloop(num,den);
w=500;
t=[0:0.001:1];
u=sin(w*t);
lsim(n,d,u,t)
```

Input frequency is
$\omega = 500$ rad/sec.

FIGURE 8.8
Output response of the Ambler closed-loop system when the input is a sinusoidal
signal of frequency $\omega = 500$ rad/sec.

shoot specifications for the Ambler mobile robot. Using simple analytic methods, we obtained an initial set of controller parameters. The controller thus designed proved to nearly satisfy the design requirements. Some fine-tuning would
be necessary to meet the design specifications exactly.

E X E R C I S E S

E8.1 Consider the mass-spring-damper system described in Section 8.2. Suppose that
we have the following parameter values: $m = 1, k = 10$, and $b = 0.1$. Also, sup-

pose that the input is a sinusoidal signal. Develop a simulation using MATLAB to study the effects on the steady-state output response of varying the input sinusoidal frequency (hint: use lsim). As part of the investigations, obtain a Bode plot of the system. At three input frequencies, verify that the steady-state magnitude of the output response correlates to the Bode plot magnitude at the same frequencies.

E8.2   Design a controller for the Ambler mobile robot of Section 8.3. Consider a controller of the form

$$G_c(s) = K\frac{s+z}{s+p},$$

where K, z and p are the key tuning parameters. Determine valid values of the tuning parameters to meet the design specifications of $P.O. < 15\%$ and $\omega_b > 1$ Hz. Can the problem be solved with this type of controller structure? Discuss the approach and methods.

E8.3   Re-consider the design problem in Section 8.5. Study the effects of increasing the parameter α. Select four values of α and catalog the following data: $P.O.$, ω_b, T_s, and $M_{p\omega}$. Based on the results, recommend a value of α.

Stability in the Frequency Domain

Hot Ingot Robot Control Example

9.1 Introduction 158

9.2 Hot Ingot Robot Control 158

9.3 Proportional Controller Design 161

9.4 Nyquist Plot for a System with a Time-delay 163

9.5 Padé Approximation 168

9.6 Other Time-delay Approximations 170

9.7 Nyquist Plot with Padé Approximation 171

9.8 PI Controller Design 173

9.9 Summary 178

   Exercises 179

PREVIEW

We use the hot ingot robot control problem to illustrate the issue of stability in the frequency-domain. Proportional and proportional-integral controllers are considered as possible options for the closed-loop system control. Also we analyze the effect of a time-delay on the system stability. As a means for the designer to include rational function approximations of the time-delay in the design process, we present approximate methods, namely Padé approximations. We also discuss the effect of time-delays on the Nyquist plot.

9.1   INTRODUCTION

The notion of **stability** is fundamental to feedback control system design. Just as there are many methods and techniques to design control laws, there are similarly many tools available to investigate system stability. In Chapter 6 of *Modern Control Systems* we considered the Routh-Hurwitz method; in Chapter 7 we considered root locus. Both methods allow us to evaluate stability, and we can use this knowledge in the design process.

In this chapter we discuss the issue of stability from a frequency-domain perspective. The Bode plot is yet another tool we can use to investigate stability by **phase** and **gain margins**. Also in this chapter we present the Nyquist criteria. Which method we use in the design process depends on the problem at hand and personal preference. Design specifications may include a mixture of time-domain and frequency-domain specifications. The hot ingot robot example will give us the opportunity to address the situation of mixed specification types: percent overshoot (time-domain) and phase margin (frequency-domain). The design elements emphasized in this chapter are shown in Figure 9.1.

9.2   HOT INGOT ROBOT CONTROL

The hot ingot robot mechanism is shown in Figure 9.2. This problem is adapted from DP9.10 in *Modern Control Systems*. The robot picks up hot ingots and sets them in a quenching tank. A vision sensor is in place to provide a measurement of the ingot position. The controller uses the sensed position information to orient the robot over the ingot (along the x-axis). The vision sensor provides the desired position input, $R(s)$, to the controller. The block diagram depiction of the closed-loop system is shown in Figure 9.3. More information on robots and robot vision systems can be found in [58]-[60].

The position of the robot along the track is also measured (by a sensor other than the vision sensor) and is available for feedback to the controller. We assume that the position measurement is noise free. This is not a restrictive assumption since many accurate position sensors are available today. For example some laser diode systems are self-contained (including the power supply, optics, laser dioide, and so on) and provide position accuracy of over 99.9%. All this comes packaged in a $2'' \times 0.5''$ case!

The robot dynamics are modeled as a second-order system with two poles at $s = -1$ and include a time-delay of $T = \pi/4$ seconds. Therefore,

$$G(s) = \frac{e^{-sT}}{(s+1)^2} \,, \tag{9.1}$$

where $T = \pi/4$ seconds.

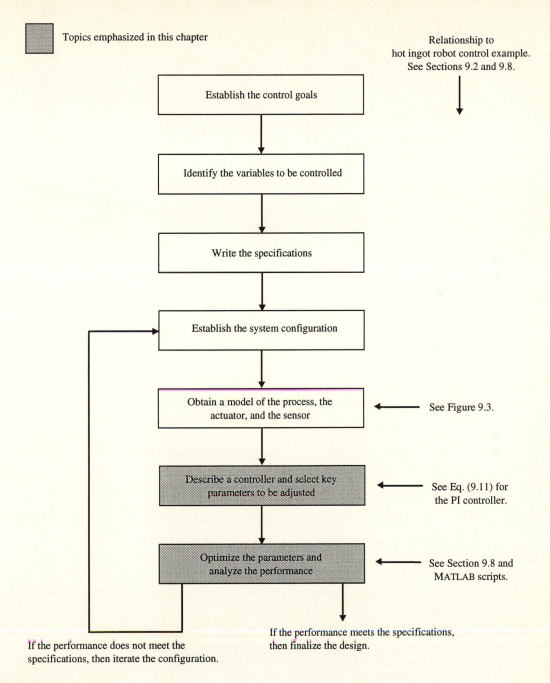

FIGURE 9.1
Elements of the control system design process emphasized in this chapter.

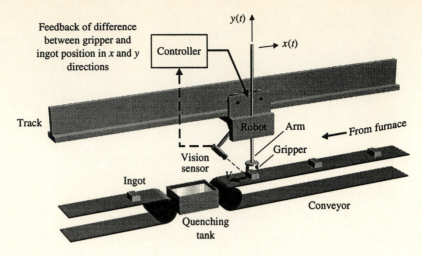

FIGURE 9.2
Artist's depiction of the hot ingot robot control system.

The control goal is as follows:

Control Goal	Minimize the tracking error, $E(s) = R(s) - Y(s)$, in the presence of external disturbances while accounting for the known time-delay.

To this end the following control specifications must be satisfied:

Design Specifications	**DS1** Achieve a steady-state tracking error less than 10% for a step input.
	DS2 Phase margin greater than 60° with the time-delay $T = \pi/4$ seconds.
	DS3 Percent overshoot less than 10% for a step input.

Our design method is first to consider a proportional controller. We will

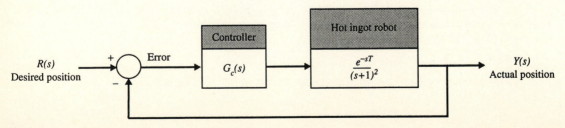

FIGURE 9.3
Hot ingot robot control system block diagram.

show that the design specifications cannot be simultaneously satisfied with a proportional controller; however, the feedback system with proportional control provides a useful vehicle to discuss in some detail the effects of the time-delay. In particular, we consider the effects of the time-delay on the Nyquist plot. The final design uses a PI controller, which is capable of providing adequate performance (that is, it satisfies all design specifications).

9.3   PROPORTIONAL CONTROLLER DESIGN

As a first try, we consider a simple proportional controller:

$$G_c(s) = K .$$

Then ignoring the time-delay for the moment, we have

$$G_c(s)G(s) = \frac{K}{(s+1)^2} = \frac{K}{s^2 + 2s + 1} .$$

The feedback control system is shown in Figure 9.4 with a proportional controller and no time-delay. The system is a type zero system, so we expect a nonzero steady-state tracking error to a step input (see Chapter 5 in *Modern Control Systems* for a review of system type). The closed-loop transfer function is

$$T(s) = \frac{K}{s^2 + 2s + 1 + K} .$$

The tracking error is defined as

$$E(s) = R(s) - Y(s) .$$

Therefore

$$E(s) = (1 - T(s))R(s) .$$

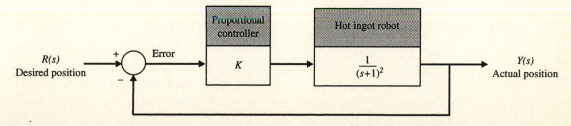

FIGURE 9.4
Hot ingot robot control system block diagram with the proportional controller and no time-delay.

But with $R(s) = a/s$, where a is the input magnitude, we have

$$E(s) = \frac{s^2 + 2s + 1}{s^2 + 2s + 1 + K} \, \frac{a}{s} \,.$$

Using the **final value theorem** (which is possible since the system is stable for all positive values of K) yields

$$e_{ss} = \lim_{s \to 0} s E(s) = \frac{a}{1 + K} \,.$$

Per specification DS1, we require the steady-state tracking error be less than 10%. Therefore

$$e_{ss} \leq \frac{a}{10} \,.$$

Solving for the appropriate gain, K, yields

$$K \geq 9 \,.$$

With $K = 9$ we obtain the Bode plot shown in Figure 9.5. Notice that we can generate the Bode plot with the margin function. In Chapter 9 of *Modern Control Systems* we state that the margin function can be used only after the bode function. With MATLAB 4.2 we can use the margin function by itself, as shown in Figure 9.5.

If we raise the gain above $K = 9$, we find that the crossover moves to the right (that is, ω_c increases) and the corresponding phase margin ($P.M.$) decreases. Is a $P.M. = 38.94°$ at $\omega = 2.828$ rad/sec sufficient for stability in the presence of a time-delay of $T = \pi/4$ seconds? The addition of the time-delay term causes a phase lag without changing the magnitude plot. The amount of time-delay that our system can withstand while remaining stable is

$$\phi = -\omega T \quad \Rightarrow \quad \frac{-38.94\pi}{180} = -2.828T \,.$$

Solving for T yields

$$T = 0.24 \text{ seconds} \,.$$

Thus for time-delays less than $T = 0.24$ seconds, our closed-loop system remains stable. However our time-delay $T = \pi/4$ seconds will cause instability. Raising the gain only exacerbates matters, since the $P.M.$ goes down further. Lowering the gain raises the $P.M.$, but the steady-state tracking error exceeds the 10% limit. Obviously, a more complex controller is necessary. Before proceeding, let us consider the Nyquist plot and see how it changes with the addition of the time-delay.

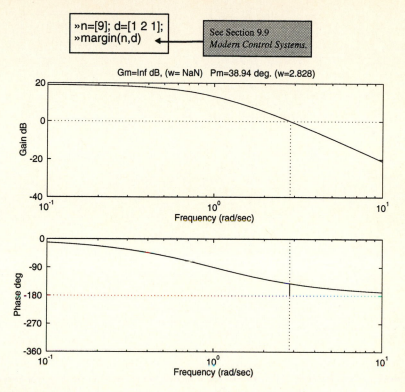

FIGURE 9.5
Bode plot with $K = 9$ and no time-delay showing gain margin $G.M. = \infty$ and phase margin $P.M. = 38.94°$.

9.4 NYQUIST PLOT FOR A SYSTEM WITH A TIME-DELAY

The Nyquist plot for the system (without the time-delay)

$$G_c(s)G(s) = \frac{K}{(s+1)^2},$$

is shown in Figure 9.6, where we use $K = 9$ (based on the discussion in Section 9.2). The number of open-loop poles of $G_c(s)G(s)$ in the right half-plane is

$$P = 0.$$

From Figure 9.6 we see that there are no encirclements of the minus 1 point, thus

$$N = 0.$$

By the **Nyquist theorem**, we know that the net number of encirclements, N, equals the number of zeros, Z, (or closed-loop system poles) in the right half-

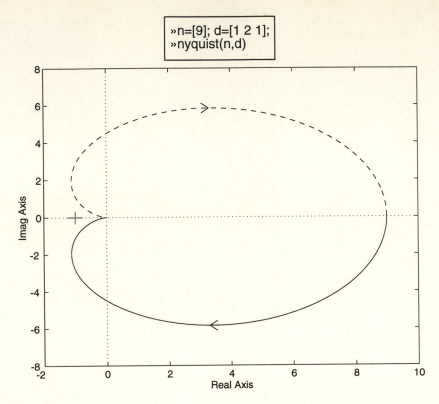

FIGURE 9.6
Nyquist plot with $K = 9$ and no time-delay showing no encirclements of the minus 1 point.

plane minus the number of open-loop poles, P, in the right half-plane. Therefore

$$Z = N + P = 0 \, .$$

Since $Z = 0$, the closed-loop system is stable. More importantly even when the gain, K, is increased (or decreased), the minus 1 point is never encircled—the gain margin is ∞. Similarly when the time-delay is absent, the phase margin is always positive. The value of the $P.M.$ varies as K varies, but the $P.M.$ is always greater than zero.

With the time-delay in the loop, we can rely on analytic methods to obtain the Nyquist plot. As we will soon see, by using the Padé approximation to the time-delay we can use the MATLAB nyquist function to obtain the Nyquist plot.

The open-loop transfer function with the time-delay is

$$G_c(s)G(s) = \frac{K}{(s+1)^2} e^{-sT} \, .$$

Using the Euler identity

$$e^{-j\omega T} = \cos \omega T - j \sin \omega T \ ,$$

and substituting $s = j\omega$ into $G_c(s)G(s)$ yields

$$G_c(j\omega)G(j\omega) = \frac{K}{(j\omega + 1)^2}e^{-j\omega T} \tag{9.2}$$

$$= \frac{K}{\Delta}\left([(1 - \omega^2)\cos \omega T - 2\omega \sin \omega T\right.$$

$$\left. - j[(1 - \omega^2)\sin \omega T + 2\omega \cos \omega T]\right) \ ,$$

where

$$\Delta = (1 - \omega^2)^2 + 4\omega^2 \ .$$

Generating a plot of $\text{Re}(G_c(j\omega)G(j\omega))$ versus $\text{Im}(G_c(j\omega)G(j\omega))$ for various values of ω leads to the plot shown in Figure 9.7. With $K = 9$, the number of encirclements of the minus 1 point is

$$N = 2 \ .$$

Therefore the system is unstable since

$$Z = N + P = 2 \ .$$

Figure 9.8 shows the Nyquist plot for four values of time-delay: $T = 0$, 0.1, 0.24, and $\pi/4$=0.78 seconds. For $T = 0$ there is no possibility of an encirclement of the minus 1 point as K varies (see the upper left corner of Figure 9.8). We have stability (that is, $N = 0$) for $T = 0.1$ seconds (upper right corner), marginal stability for $T = 0.24$ seconds (lower left corner), and for $T = \pi/4 = 0.78$ seconds we have $N = 1$ (lower right corner), thus the closed-loop system is unstable.

Since we know that $T = \pi/4$ for our problem, the proportional gain controller is not a viable controller. With it we cannot meet the steady-state error specifications and have a stable closed-loop system in the presence of the time-delay $T = \pi/4$. However before proceeding with the design of a controller that meets all the specifications, let us take a closer look at the Nyquist plot with a time-delay.

Suppose we have the case where $K = 9$ and $T = 0.1$ seconds. The associated Nyquist plot is shown in the upper right corner of Figure 9.8. A close-up of the origin of the Nyquist plot is shown in Figure 9.9. Notice that the Nyquist plot spirals toward the origin. The Nyquist plot intersects (or crosses over) the real axis whenever the imaginary part of $G_c(j\omega)G(j\omega) = 0$ [see Eq. (9.2)], or

$$(1 - \omega^2)\sin 0.1\omega + 2\omega \cos 0.1\omega = 0 \ .$$

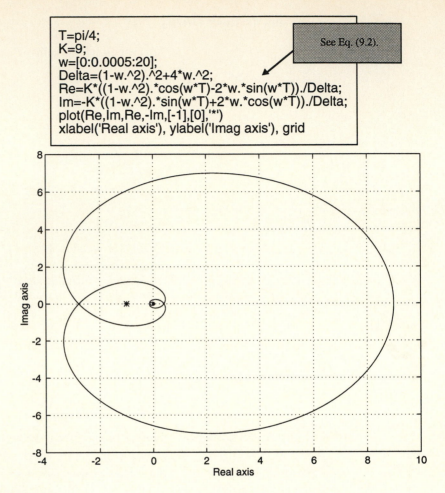

```
T=pi/4;
K=9;
w=[0:0.0005:20];
Delta=(1-w.^2).^2+4*w.^2;
Re=K*((1-w.^2).*cos(w*T)-2*w.*sin(w*T))./Delta;
Im=-K*((1-w.^2).*sin(w*T)+2*w.*cos(w*T))./Delta;
plot(Re,Im,Re,-Im,[-1],[0],'*')
xlabel('Real axis'), ylabel('Imag axis'), grid
```

See Eq. (9.2).

FIGURE 9.7
Nyquist plot with $K = 9$ and $T = \pi/4$ showing one encirclement of the minus 1 point, $N = 0$.

Thus we obtain the relation that describes the frequencies ω at which crossover occurs:

$$\frac{(1 - \omega^2)\tan 0.1\omega}{2\omega} = -1 . \tag{9.3}$$

Eq. (9.3) has an infinite number of solutions, corresponding to the spiral intersections of the real axis. The first real-axis crossing (farthest in the left half-plane) occurs when

$$\omega = 4.4352 \text{ rad/sec} .$$

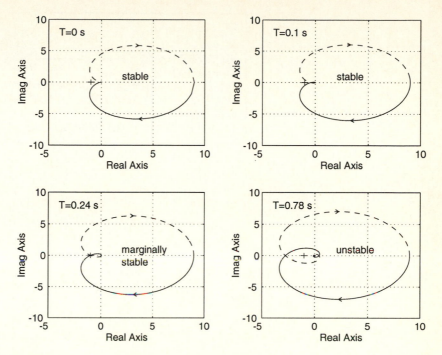

FIGURE 9.8
Nyquist plot with $K = 9$ and various time-delays.

We can obtain this result by plotting the function

$$f(\omega) = \frac{(1 - \omega^2)\tan 0.1\omega}{2\omega} + 1 \;, \tag{9.4}$$

and determining the ω associated with the first zero crossing.

The magnitude of $|G_c(j4.4352)G(j4.4352)|$ is equal to 0.0484 K. For stability we require that $|G_c(j\omega)G(j\omega)| < 1$ when $\omega = 4.4352$ (to avoid an encirclement of the minus 1 point). Thus for stability we find

$$K < \frac{1}{0.0484} = 20.67 \;,$$

when $T = 0.1$. When $K = 9$, the closed-loop system is stable, as we already know. If the gain $K = 9$ increases by a factor of $\delta = 2.3$ to $K = 20.67$, we will be on the border of instability. This factor δ is the gain margin:

$$G.M. = 20\log_{10}\delta = 20\log_{10} 2.3 = 7.2\,dB \;.$$

Verify this using the margin function.

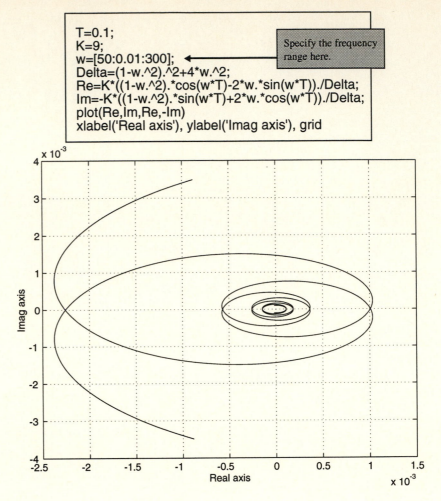

FIGURE 9.9
Close-up of region around the origin of the Nyquist plot with $K = 9$ and $T = 0.1$.

9.5 PADÉ APPROXIMATION

The systems considered by most analytical tools (root locus, Routh-Hurwitz, and so on) are described by rational functions, (that is, transfer functions) or by a finite set of ordinary constant coefficient differential equations. Some real systems, most notably those systems with a time-delay, are nonrational. Recall that the time-delay is given by e^{-sT}, where T is the delay. It would be helpful if we could obtain a rational function approximation of the time-delay. Then it would be more convenient to incorporate the delay into the block diagram for analysis and design purposes.

The **Padé** approximation uses a series expansion of the transcendental function e^{-sT} and matches as many coefficients as possible with a series expansion of a rational function of specified order. For example, to approximate the function e^{-sT} with a first-order rational function, we begin by expanding both functions in a series (actually a McLauren series[1]),

$$e^{-sT} = 1 - sT + \frac{(sT)^2}{2!} - \frac{(sT)^3}{3!} + \frac{(sT)^4}{4!} - \frac{(sT)^5}{5!} + \cdots , \qquad (9.5)$$

and

$$\frac{n_1 s + n_o}{d_1 s + d_o} = \frac{n_o}{d_o} + \left(\frac{d_o n_1 - n_o d_1}{d_o^2} \right) s + \left(\frac{d_1^2 n_o}{d_o^3} - \frac{d_1 n_1}{d_o^2} \right) s^2$$

$$+ \left(\frac{d_1^2 n_1}{d_o^3} - \frac{d_1^3 n_o}{d_o^4} \right) s^3 + \cdots . \qquad (9.6)$$

We want to find n_o, n_1, d_o, and d_1 such that

$$e^{-sT} \approx \frac{n_1 s + n_o}{d_1 s + d_o} .$$

Equating the corresponding coefficients of the terms in s, we obtain the relationships

$$\frac{n_o}{d_o} = 1 ,$$

$$\frac{n_1}{d_o} - \frac{n_o d_1}{d_o^2} = -T ,$$

$$\frac{d_1^2 n_o}{d_o^3} - \frac{d_1 n_1}{d_o^2} = \frac{T^2}{2} ,$$

$$\frac{d_1^2 n_1}{d_o^3} - \frac{d_1^3 n_o}{d_o^4} = -\frac{T_3}{6} ,$$

and so on. Using the first three equations and solving for n_o, d_o, n_1, and d_1 yields

$$n_o = d_o ,$$

$$d_1 = \frac{d_o T}{2} ,$$

$$n_1 = -\frac{d_o T}{2} .$$

---

1. $f(s) = f(0) + \frac{s}{1!} \dot{f}(0) + \frac{s^2}{2!} \ddot{f}(0) + \cdots$

So using the relationships for n_o, d_1, n_1, and d_o, we have that

$$\frac{n_1 s + n_o}{d_1 s + d_o} = \frac{-\frac{T}{2}s + 1}{\frac{T}{2}s + 1} \,. \tag{9.7}$$

A series expansion of Eq. (9.7) yields

$$\frac{n_1 s + n_o}{d_1 s + d_o} = \frac{-\frac{T}{2}s + 1}{\frac{T}{2}s + 1} = 1 - Ts + \frac{T^2 s^2}{2} - \frac{T^3 s^3}{4} + \cdots \,. \tag{9.8}$$

Comparing Eq. (9.8) to Eq. (9.5), we see that the first three terms match exactly. So for small s, our Padé approximation is a reasonable representation of the time-delay. Higher-order rational functions can be used if necessary. MATLAB provides a Padé approximation of specified order using the pade function (see Chapter 8 in *Modern Control Systems*).

9.6   OTHER TIME-DELAY APPROXIMATIONS

It is also possible to utilize the series expansion of e^{-sT} to obtain an approximation for the time-delay. We start with the series expansion:

$$e^{-sT} = 1 - Ts + \frac{T^2 s^2}{2!} - \frac{T^3 s^3}{3!} + \cdots \,.$$

Then we can truncate the series keeping as many terms as we desire. For example, keeping three terms yields

$$e^{-sT} \approx 1 - Ts + \frac{T^2 s^2}{2!} \approx \frac{1}{1 + Ts + \frac{T^2 s^2}{2}} \,. \tag{9.9}$$

For small time-delays, we can see in Eq. (9.9) that

$$e^{-sT} \approx \frac{1}{1 + Ts} \,. \tag{9.10}$$

It is common to see a first-order system of the form in Eq. (9.10) and second-order systems of the form in Eq. (9.9) used as approximations for a time-delay. In general, the Padé method provides a better approximation.

9.7   NYQUIST PLOT WITH PADÉ APPROXIMATION

We can use a Padé approximation of the time-delay to obtain a Nyquist plot for

$$G_c(s)G(s) = \frac{9e^{-sT}}{(s+1)^2} \approx \frac{9}{(s+1)^2} \, \frac{-\frac{T}{2}s+1}{\frac{T}{2}s+1} \,,$$

where $T = \pi/4$. The Nyquist plot is shown in Figure 9.10.

In Figure 9.11 Nyquist plots are generated for first- and second-order Padé approximations. When the frequency is small (that is, for small ω) the Padé ap-

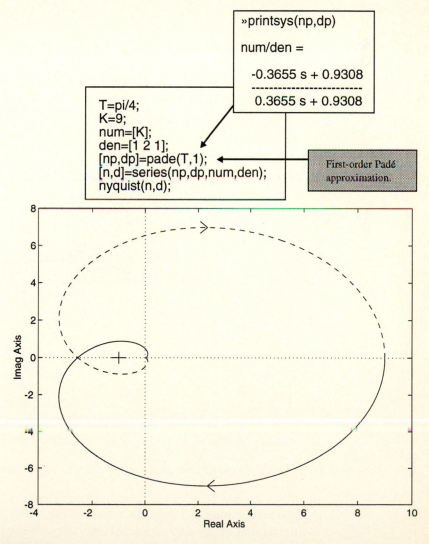

FIGURE 9.10
Nyquist plot generated by the pade function.

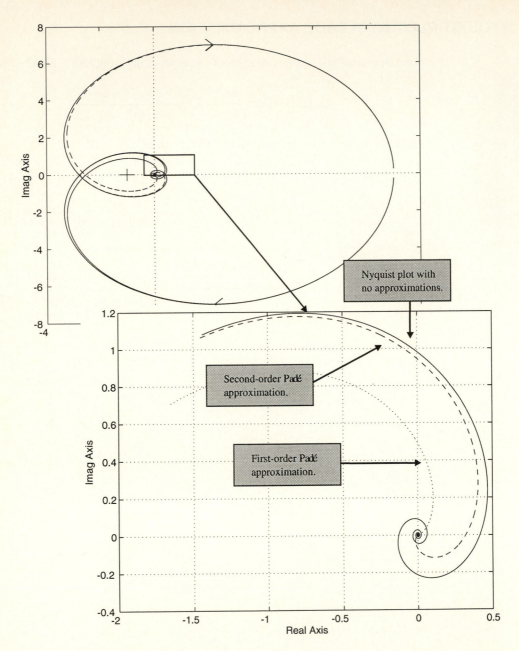

FIGURE 9.11
Close-up of the Nyquist plots generated by first- and second-order pade approximations.

proximations match closely the actual Nyquist plot. As ω increases, the plots with the time-delay approximations deviate more from the actual Nyquist. Increasing the order of the Padé approximation helps, and the order of the approximation necessary for a given problem depends on the frequencies of interest.

9.8   PI CONTROLLER DESIGN

The issue of steady-state error can be addressed by increasing the system type. Consider the PI controller

$$G_c(s) = K_P + \frac{K_I}{s} = \frac{K_P s + K_I}{s} \, . \tag{9.11}$$

The open-loop system transfer function is

$$G_c(s)G(s) = \frac{K_P s + K_I}{s} \frac{K}{(s+1)^2} e^{-sT} \, .$$

The system type is now equal to 1; thus we expect a zero steady-state error to a step input. The steady-state error specification DS1 is satisfied. We can now concentrate on meeting specification DS3, $P.O. < 10\%$ and DS2, the requirement for stability in the presence of the time-delay $T = \pi/4$ seconds.

From the percent overshoot specification we can determine a desired system damping ratio. Thus we determine for $P.O. \leq 10\%$ that

$$\zeta \geq 0.59 \, .$$

This follows from the design formula

$$P.O.(\%) = 100 e^{-\frac{\pi\zeta}{\sqrt{1-\zeta^2}}} \, ,$$

for second-order systems. Our system is third-order, however, so this is an approximation (a valuable approximation nonetheless). Due to the PI controller, the system now has a zero at $s = -K_I/K_P$. The zero will not affect the closed-loop system stability, but it will affect the performance. Using the approximation (valid for small ζ)

$$\zeta \approx \frac{P.M.}{100} \, ,$$

we determine a good target phase margin (since we want $\zeta \geq 0.59$) to be 60%. We can rewrite the PI controller as

$$G_c(s) = K_I \frac{1 + \tau s}{s} \, ,$$

where $1/\tau = K_I/K_P$ is the break frequency of the controller. The PI controller is essentially a low-pass filter and adds phase lag to the system below the break frequency. We would like to place the break frequency below the crossover frequency so that the phase margin is not reduced significantly due to the presence of the PI zero.

The uncompensated Bode plot is shown in Figure 9.12 for

$$G(s) = \frac{9}{(s+1)^2}e^{-sT} \, ,$$

where $T = \pi/4$. We use a sixth-order Padé approximation to model the time-delay. The uncompensated system phase margin is

$$P.M. = -88.34°$$

at $\omega_c = 2.83$ rad/sec. Since we want $P.M. = 60°$, we need the phase to be

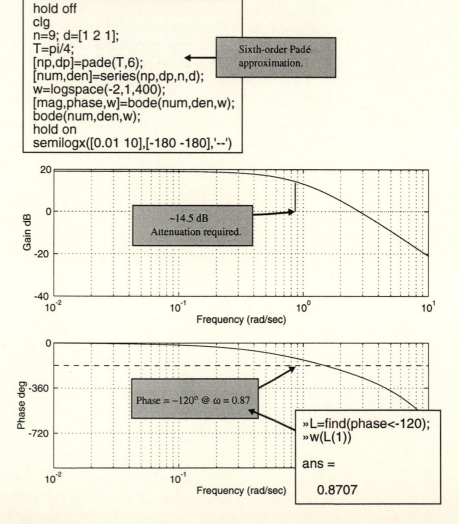

FIGURE 9.12
Uncompensated Bode plot with $K = 9$ and $T = \pi/4$.

minus 120° at the crossover frequency. In Figure 9.12 we can estimate the phase

$$\phi = -120°$$

at $\omega \approx 0.87$ rad/sec. Using MATLAB we can estimate this by searching the phase variable for the first occurence of phase less than $-120°$, as shown in Figure 9.12. This is an approximate value but is sufficiently accurate for the design procedure. At $\omega = 0.87$ the magnitude is about 14.5 dB. If we want the crossover to be $\omega_c = 0.87$ rad/sec, the controller needs to attenuate the system gain by 14.5 dB, so that the magnitude is 0 dB at $\omega_c = 0.87$. With

$$G_c(s) = K_P \left(\frac{s + \frac{K_I}{K_P}}{s} \right) ,$$

we can consider K_P to be the gain of the compensator (a good approximation for large ω). Therefore,

$$K_P = 10^{-(14.5/20)} = 0.188 .$$

Finally we need to select K_I. Since we want the break frequency of the controller to be below the crossover frequency (so that the phase margin is not reduced significantly due to the presence of the PI zero), a good rule-of-thumb is to select

$$1/\tau = K_I/K_P = 0.1\omega_c .$$

The break frequency of the controller's zero is one decade below the crossover frequency. The final value of K_I is computed to be

$$K_I = 0.1\omega_c K_P = 0.0164 ,$$

where $\omega_c = 0.87$ rad/sec. Thus the PI controller is

$$G_c(s) = \left(\frac{0.188s + 0.0164}{s} \right) . \tag{9.12}$$

The Bode plot of $G_c(s)G(s)$ is shown in Figure 9.13. The gain and phase margins are

$$G.M. = 5.3 \text{ dB} ,$$
$$P.M. = 56.5° .$$

Have the design specifications been met? The steady-state tracking specification (DS1) is certainly satisfied since our system is type one; the PI controller

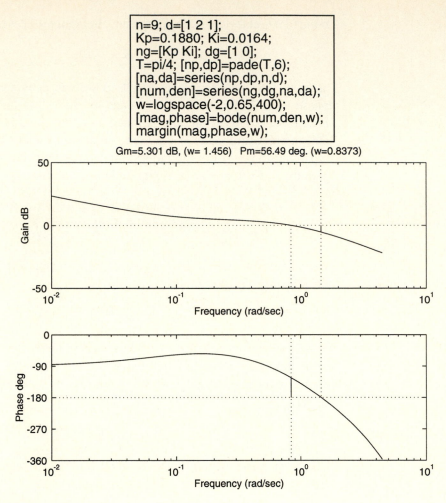

FIGURE 9.13
Compensated Bode plot with $K = 9$ and $T = \pi/4$.

introduced an integrator. The phase margin (with the time-delay) is

$$P.M. = 56.5° \,.$$

Some iteration can be performed to increase the phase margin to the target of 60°. Raising the phase margin will tend to increase the settling time, but this should be an acceptable trade-off since the design specification list does not include a settling time requirement.

The Nichols diagram is shown in Figure 9.14. The phase margin and gain margin can be obtained directly from the diagram. The phase is found to be

$$\phi \approx -123.5°$$

when the magnitude is 0 dB. Thus the phase margin is

$$P.M. = 180 - 123.5 = 56.5° .$$

The unit step response is shown in Figure 9.15. The percent overshoot is approximately

$$P.O. \approx 4.2\% .$$

The target percent overshoot was 10%, so DS3 is satisfied. Overall the design specifications were satisfied or very nearly satisfied. Some iteration on the controller design would be necessary to meet the phase margin specification of 60°.

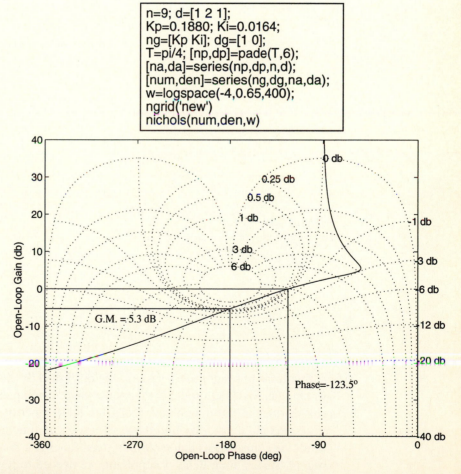

FIGURE 9.14
Nichols chart of the system with the PI controller.

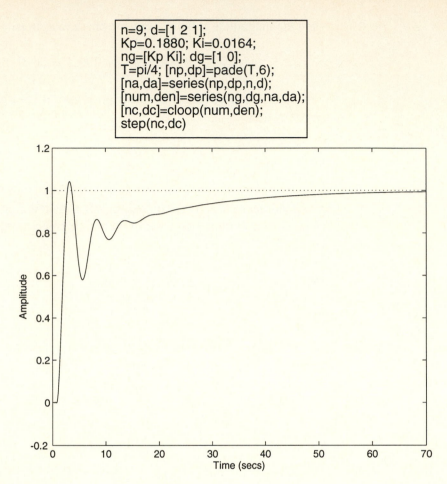

```
n=9; d=[1 2 1];
Kp=0.1880; Ki=0.0164;
ng=[Kp Ki]; dg=[1 0];
T=pi/4; [np,dp]=pade(T,6);
[na,da]=series(np,dp,n,d);
[num,den]=series(ng,dg,na,da);
[nc,dc]=cloop(num,den);
step(nc,dc)
```

FIGURE 9.15
Hot ingot robot control step response.

9.9   SUMMARY

In this chapter we designed a PI controller to meet the specifications associated with the hot ingot robot system. We discussed the effect of time-delays on
system stability (and on the Nyquist plot) and developed transfer function approximations to the time-delay. Using the Padé function in MATLAB, we then
introduced the time-delay into the system transfer function. This allowed us to
use all the control tools that require rational functions (that is, transfer function)
models. Since a time-delay was present, we needed to lower the system gain to
improve the phase margin, but in doing so we reduced the steady-state tracking
performance.

E X E R C I S E S

E9.1   What if the position sensor, which is normally noise-free, begins to fail in such a manner that its measurements become very noisy? Suppose that the noise is in the frequency range $\omega > 100$ Hz. Study the effects of the noise on the control system with the PI controller as designed ($K_P = 0.188$ and $K_I = 0.0164$).

E9.2   Redesign the PI controller in Section 9.8 using a controller of the form

$$G_c(s) = \frac{s + a}{s + b}.$$

Is there anything gained from switching to this controller? Discuss the results.

CHAPTER 10

Design of Feedback Control Systems

Milling Machine Control Example

10.1 Introduction   180
10.2 Lead and Lag Compensators   181
10.3 Milling Machine Control System   181
10.4 Lag Compensator Design   187
10.5 Summary   191
     Exercises   191

PREVIEW

This chapter begins with a review of lead and lag compensators. The two compensators have the same general structure but differ in the relative locations of their poles and zeros. We use a milling machine example to illustrate the design of a lag compensator using root locus methods. We obtain the milling machine model by appropriately fitting the impulse response data of the milling machine. The lag compensator adequately satisfies the design specifications.

10.1   INTRODUCTION

The main theme of this chapter is design of feedback control systems; design has been the main topic of all the previous chapters as well. There are many aspects of the design process, and each previous chapter emphasized different elements of the process. The design process elements emphasized in this chapter are shown in Figure 10.1.

In this chapter we focus on the **lead** and **lag compensators**. Previously we worked almost exclusively with PID controllers and two variations—PI and PD controllers. The PID is an important controller structure, but the lead and lag compensators are also frequently used in engineering practice.

10.2   LEAD AND LAG COMPENSATORS

A good introduction to lead and lag compensators can be found in Chapter 10 of *Modern Control Systems*. A very brief review of that material is given here.

Consider a first-order compensator of the form

$$G_c(s) = K \frac{s + z}{s + p} . \tag{10.1}$$

Both the lead and lag compensators have the form given in Eq. (10.1). The difference lies in the selection of the zero at $s = -z$ and the pole at $s = -p$. When

$$|z| < |p| ,$$

we have a lead compensator. Conversely when

$$|p| < |z| ,$$

we have a lag compensator. The lead and lag compensators are summarized in Figure 10.2 and Figure 10.3, respectively.

10.3   MILLING MACHINE CONTROL SYSTEM

Smaller, lighter, less costly sensors are being developed by engineers for machining and other manufacturing processes. A milling machine table is depicted in Figure 10.4. The milling machine control system design example of this section is adapted from P10.36 in *Modern Control Systems*.

This particular machine table has a new sensor that gleans information about the cutting process (that is, the depth-of-cut) from the acoustic emission (AE) signals. **Acoustic emissions** are low-amplitude, high-frequency stress waves that originate from the rapid release of strain energy in a continuous medium. The AE sensors are commonly piezoelectric amplitude sensitive in the 100 kHz to 1 MHz range; they are cost effective and can be mounted on most machine tools.

There is a relationship between the sensitivity of the AE power signal and small depth-of-cut changes ([61]–[63]). This relationship can be exploited to

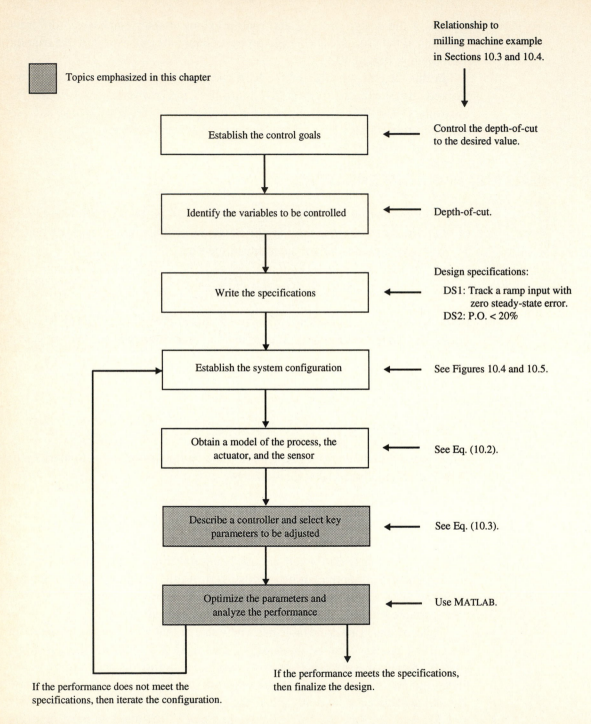

FIGURE 10.1
Elements of the control system design process emphasized in this chapter.

$$G_c(s) = K \frac{s+z}{s+p} \ , \ |z| < |p|$$

$$G_c(s) = \frac{K}{\alpha} \frac{1+\alpha\tau s}{1+\tau s} \ , \ \tau = \frac{1}{p} \ , \ \alpha = \frac{p}{z} > 1$$

a) Phase-lead transfer functions

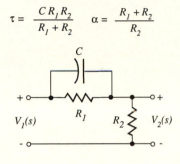

b) Phase-lead pole-zero map

$$\tau = \frac{C R_1 R_2}{R_1 + R_2} \qquad \alpha = \frac{R_1 + R_2}{R_2}$$

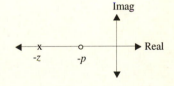

c) Phase-lead network

d) Bode diagram of phase-lead network.

FIGURE 10.2
Summary of the lead compensator.

$$G_c(s) = \frac{K}{\alpha} \frac{s+z}{s+p} \ , \ |p| < |z|$$

$$G_c(s) = \frac{1+\tau s}{1+\alpha\tau s} \ , \ \tau = \frac{1}{z} \ , \ \alpha = \frac{z}{p} > 1$$

a) Phase-lag transfer functions

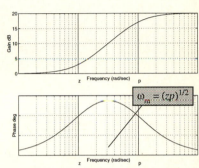

b) Phase-lag pole-zero map

$$\tau = C R_2 \qquad \alpha = \frac{R_1 + R_2}{R_2}$$

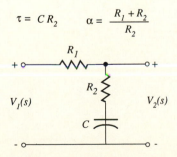

c) Phase-lag network

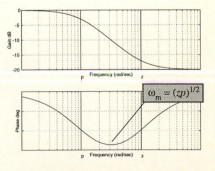

d) Bode diagram of phase-lag network

FIGURE 10.3
Summary of the lag compensator.

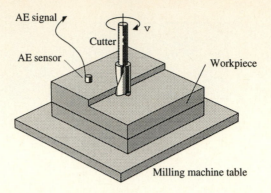

FIGURE 10.4
An artist's depiction of the milling machine.

obtain a feedback signal or measurement of the depth-of-cut. A simplified block diagram of the feedback system is shown in Figure 10.5.

Since the acoustic emissions are sensitive to material, tool geometry, tool wear, and cutting parameters such as cutter rotational speed, the measurement of the depth-of-cut is modeled as being corrupted by noise, denoted by $N(s)$ in Figure 10.5. Also disturbances on the plant, denoted by $D(s)$, are modeled. These might represent external disturbances resulting in unwanted motion of the cutter, fluctuations in the cutter rotation speed, and so forth.

The plant model, $G(s)$, is given by

$$G(s) = \frac{2}{s(s+1)(s+5)} , \qquad (10.2)$$

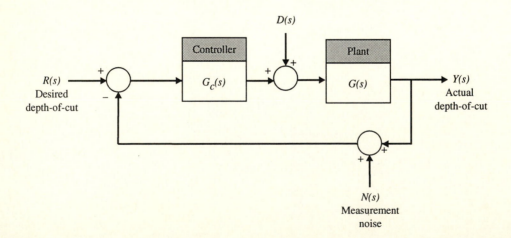

FIGURE 10.5
A simplified block diagram of the milling machine feedback system.

and represents the model of the cutter apparatus and the AE sensor dynamics. The input to $G(s)$ is a control signal to actuate an electromechanical device, which then applies downward pressure on the cutter.

There are a variety of methods available to obtain the model in Eq. (10.2). One approach would be to use basic principles to obtain a mathematical model in the form of a nonlinear differential equation, which can then be linearized about an operating point leading to a linear model (or equivalently, a transfer function). The basic principles include Newton's laws, the various conservation laws, and Kirchhoff's laws.

Another approach would be to assume a form of the model (such as, a second-order system) with unknown parameters (such as, ω_n and ζ), and then experimentally obtain good values of the unknown parameters.

A third approach is to conduct a laboratory experiment to obtain the step or impulse response of the system. In other words we can apply an input (in this case, a voltage) to the system and measure the output—the depth-of-cut into the desired workpiece. Suppose, for example, we have the impulse response data shown in Figure 10.6 (the small circles on the graph represent the data).

If we had access to the function $C_{\text{imp}}(t)$—the impulse response function of the milling machine — we could take the Laplace transform to obtain the transfer function model. There are many methods available for **curve-fitting** the

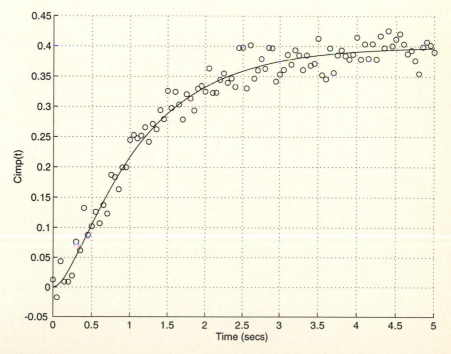

FIGURE 10.6
Hypothetical impulse response of the milling machine.

data to obtain the function $C_{\text{imp}}(t)$. We will not cover curve-fitting here, but we can say a few words regarding the basic structure of the function.

From Figure 10.6 we see that the response approaches a steady-state value:

$$C_{\text{imp}}(t) \rightarrow C_{\text{imp,ss}} \approx \frac{2}{5} \text{ as } t \rightarrow \infty .$$

So we expect that

$$C_{\text{imp}}(t) = \frac{2}{5} + \Delta C_{\text{imp}}(t) ,$$

where $\Delta C_{imp}(t)$ is a function that goes to zero as t gets large. This leads us to consider $\Delta C_{\text{imp}}(t)$ as a sum of stable exponentials. Since the response does not oscillate, we might expect that the exponentials are, in fact, real exponentials,

$$\Delta C_{\text{imp}}(t) = \sum_i k_i e^{-\tau_i t} ,$$

where τ_i are positive real numbers. The data in Figure 10.6 (which are not actual experimental data) are fit by the function

$$C_{\text{imp}}(t) = \frac{2}{5} + \frac{1}{10}e^{-5t} - \frac{1}{2}e^{-t} ,$$

and the Laplace transform is

$$G(s) = \mathcal{L}\{C_{\text{imp}}(t)\} = \frac{2}{5}\frac{1}{s} + \frac{1}{10}\frac{1}{s+5} - \frac{1}{2}\frac{1}{s+1} = \frac{2}{s(s+1)(s+5)} .$$

Thus we can obtain the transfer function model of the milling machine.

The control goal is to develop a feedback system to track a desired step input. In this case the reference input is the desired depth-of-cut. The control goal is stated as

---

Control Goal Control the depth-of-cut to the desired value.

The variable to be controlled is the depth-of-cut, or

Variable to Be Controlled Depth-of-cut, $y(t)$.

Since we are focusing on lead and lag controllers in this chapter, the key tuning parameters are the parameters associated with the compensator given in Eq. (10.1).

Select Key Tuning Parameters	Compensator variables: p, z, and K.

The control design specifications are

Control Design Specifications	**DS1** Track a ramp input, $R(s) = a/s^2$, with a steady-state tracking error less than $a/8$, where a is the ramp magnitude.
	DS2 Percent overshoot to a step input less than 20%.

10.4   LAG COMPENSATOR DESIGN

The lag compensator (see Figure 10.3) is given by

$$G_c(s) = \frac{K}{\alpha} \frac{s+z}{s+p}, \quad |p| < |z|, \tag{10.3}$$

where $\alpha = z/p$. The tracking error is

$$E(s) = R(s) - Y(s) = (1 - T(s))R(s),$$

where

$$T(s) = \frac{G_c(s)G(s)}{1 + G_c(s)G(s)}.$$

Therefore,

$$E(s) = \frac{1}{1 + G_c(s)G(s)} R(s).$$

With $R(s) = a/s^2$ and using the final value theorem, we find that

$$e_{ss} = \lim_{t \to \infty} e(t) = \lim_{s \to 0} sE(s) = \lim_{s \to 0} s \frac{1}{1 + G_c(s)G(s)} \frac{a}{s^2},$$

or equivalently,

$$\lim_{s \to 0} sE(s) = \frac{a}{\lim_{s \to 0} sG_c(s)G(s)}.$$

According to DS1, we require that

$$\frac{a}{\lim_{s \to 0} sG_c(s)G(s)} < \frac{a}{8},$$

or

$$\lim_{s \to 0} s G_c(s) G(s) > 8.$$

Substituting for $G(s)$ and $G_c(s)$ from Eqs. (10.2) and (10.3), respectively, we obtain the compensated velocity constant

$$K_{v_{\text{comp}}} = \frac{2}{5} \left(\frac{K}{\alpha} \right) \left(\frac{z}{p} \right) = \left(\frac{2}{5} \overline{K} \right) \left(\frac{z}{p} \right) > 8,$$

where $\overline{K} = K/\alpha$. The compensated velocity constant is the velocity constant of the system when the lag compensator is in the loop.

The loop transfer function is

$$G_c(s) G(s) = \left[\frac{s+z}{s+p} \right] \left[\frac{2\overline{K}}{s(s+1)(s+5)} \right].$$

We separate the lag compensator from the plant and obtain the uncompensated root locus by considering the feedback loop with the gain $\overline{K}$, but not the lag compensator zero and pole factors. The uncompensated root locus for the characteristic equation

$$1 + \overline{K} \frac{2}{s(s+1)(s+5)} = 0$$

is shown in Figure 10.7.

From DS2 we determine that the target damping ratio of the dominant roots is $\zeta > 0.45$. Using the rlocfind function in MATLAB, we find that

$$\overline{K} = 2.48 \quad \text{at} \quad \zeta = 0.45.$$

Then with $\bar{K} = 2.48$ the **uncompensated velocity constant** is

$$K_{v_{\text{uncomp}}} = \lim_{s \to 0} s \frac{2\overline{K}}{s(s+1)(s+5)} = \frac{2\overline{K}}{5} = 0.993.$$

The **compensated velocity constant** is

$$K_{v_{\text{comp}}} = \lim_{s \to 0} s \left[\frac{s+z}{s+p} \right] \left[\frac{2\overline{K}}{s(s+1)(s+5)} \right] = \frac{z}{p} K_{v_{\text{uncomp}}}.$$

Therefore with $\alpha = z/p$, we obtain the relationship

$$\alpha = \frac{K_{v_{\text{comp}}}}{K_{v_{\text{uncomp}}}}.$$

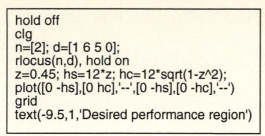

```
hold off
clg
n=[2]; d=[1 6 5 0];
rlocus(n,d), hold on
z=0.45; hs=12*z; hc=12*sqrt(1-z^2);
plot([0 -hs],[0 hc],'--',[0 -hs],[0 -hc],'--')
grid
text(-9.5,1,'Desired performance region')
```

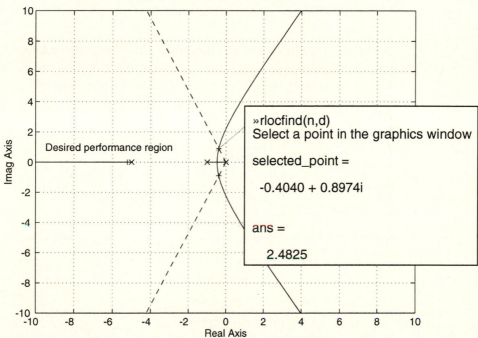

FIGURE 10.7
Root locus for the uncompensated system.

We require $K_{v_{\text{comp}}} > 8$. A possible choice is $K_{v_{\text{comp}}} = 10$ as the desired velocity constant. Then

$$\alpha = \frac{K_{v_{\text{comp}}}}{K_{v_{\text{uncomp}}}} = \frac{10}{0.993} = 10.07 \ .$$

But $\alpha = z/p$, thus our lag compensator should have $p = 0.0993z$. If we select

$$z = 0.01$$

then

$$p \approx 0.001 \ .$$

The compensated loop transfer function is given by

$$G_c(s)G(s) = \overline{K} \left[\frac{s+z}{s+p} \right] \left[\frac{2}{s(s+1)(s+5)} \right] .$$

The lag compensator with z and p as above is determined to be

$$G_c(s) = 2.48 \frac{s+0.01}{s+0.001} . \tag{10.4}$$

The step response is shown in Figure 10.8. The percent overshoot is approximately 26%. To exactly satisfy the design specification DS1, we would need to fine-tune the controller parameters. The velocity error constant is approximately 10, which satisfies DS2.

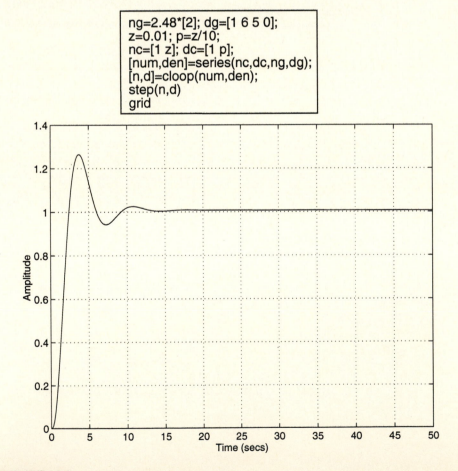

```
ng=2.48*[2]; dg=[1 6 5 0];
z=0.01; p=z/10;
nc=[1 z]; dc=[1 p];
[num,den]=series(nc,dc,ng,dg);
[n,d]=cloop(num,den);
step(n,d)
grid
```

FIGURE 10.8
Step response for the compensated system.

10.5   SUMMARY

In this chapter we reviewed the lead and lag compensators. The two compensators have the same general structure: one pole and one zero. They differ in the relative location of the pole and zero. Using the milling machine control example and root locus methods, we designed a lag compensator. We obtained the milling machine model by appropriately fitting the impulse response data of the milling machine. The lag compensator satisfied the design specifications adequately, although the percent overshoot was slightly higher than the desired 20%. Iteration of the initial lag compensator design would be necessary to obtain the final design.

E X E R C I S E S

E10.1   Rework the milling machine control system design to meet the design specifications exactly. Iterate the lag compensator parameters to meet the percent overshoot specification of 20%.

E10.2   Design a lag compensator for the milling machine example using frequency response methods. Discuss the design experience, and compare and contrast the resulting compensator to the one we developed in Section 10.4 using root locus methods.

E10.3   Design a control system using a lead compensator for the milling machine depth-of-cut. Which compensator is preferable—lead or lag—in this situation? Discuss.

E10.4   Discuss the trade-offs, advantages, and disadvantages of using PI, PD, PID, lead, and lag compensators for the milling machine problem.

CHAPTER 11

Design of State Variable Feedback Systems

Diesel Electric Locomotive Example

11.1 Introduction 193

11.2 Robot Drive Train Dynamics 195

11.3 More on Controllability 199

11.4 More on Observability 203

11.5 Diesel Electric Locomotive Example 204

11.6 State Feedback Controller Design 206

11.7 System Simulation with SIMULINK 211

11.8 Summary 214

Exercises 215

P R E V I E W

This chapter focuses on state feedback design of controllers using state variable models. We emphasize the importance of controllability and observability in state feedback control system design. A necessary and sufficient condition for arbitrarily placing the closed-loop poles using state feedback is that the system be controllable. Given a controllable system, we can use Ackermann's formula to obtain the feedback gain matrix that places the closed-loop poles at the desired locations. This approach to control system design is known as pole placement. We use a simple example based on robot drive train dynamics to illustrate the state variable modeling process. The chapter concludes with the design of a state feedback controller for a diesel electric locomotive problem. We present a SIMULINK simulation to assist in the analysis of the control design and to provide a mechanism to perform numerical experiments and answer what if questions.

11.1   INTRODUCTION

In Chapter 3 of *Modern Control Systems* and in this supplement, we discussed modeling of systems using state variable methods. We now extend the discussion to include control system design with **state variable models**. The design process elements emphasized in this chapter are shown in Figure 11.1. A state variable (or state-space) model with n state variables is represented by the matrix equations

$$\dot{\mathbf{x}} = \mathbf{A}\mathbf{x} + \mathbf{B}u \,, \tag{11.1}$$
$$y = \mathbf{C}\mathbf{x} + \mathbf{D}u \,,$$

where $\mathbf{A}$ is an $n \times n$ matrix, $\mathbf{B}$ is a $n \times 1$ matrix, $\mathbf{C}$ is a $1 \times n$ matrix, and $\mathbf{D}$ is a 1×1 matrix. More generally, the matrices $\mathbf{B}$, $\mathbf{C}$, and $\mathbf{D}$ are $n \times m$, $p \times n$, and $p \times m$, respectively, where m is the number of inputs and p is the number of outputs. We consider primarily single-input, single-output systems; therefore, $m = 1$ and $p = 1$.

The state of the system is described in terms of the state variables. The **state variables** are denoted by $x_1(t), x_2(t), \cdots, x_n(t)$. We represent the set of state variables with $\mathbf{x}(t)$ in vector form as follows:

$$\mathbf{x}(t) = \begin{bmatrix} x_1(t) \\ x_2(t) \\ \vdots \\ x_n(t) \end{bmatrix} .$$

The state variables describe the future behavior of the system when the present state of the system is known and all the future input signals, $u(t)$, are known. The dynamics of the system are given by the model in Eq. (11.1), the state variable model.

Given the state variable model, we can derive a transfer function model. The transfer function associated with Eq. (11.1) is

$$G(s) = \frac{Y(s)}{U(s)} = \mathbf{C}(s\mathbf{I} - \mathbf{A})^{-1}\mathbf{B} + \mathbf{D} \,. \tag{11.2}$$

Using the MATLAB functions tf2ss and ss2tf, we can obtain state variable models

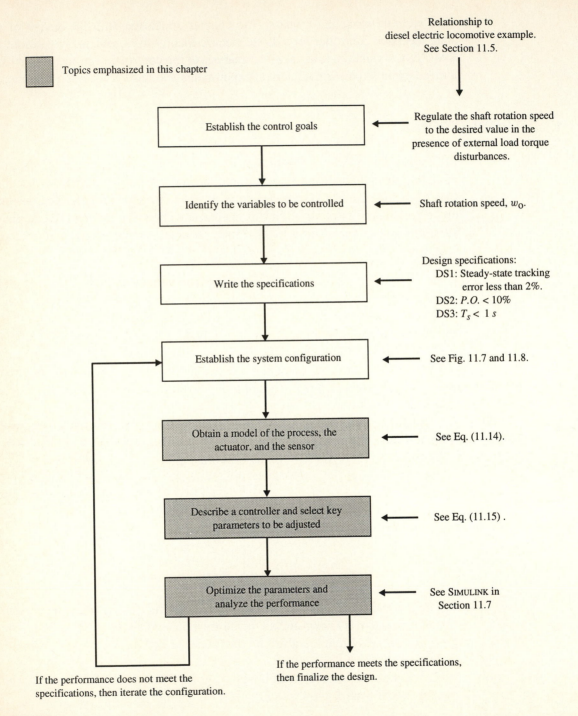

FIGURE 11.1
Elements of the control system design process emphasized in this chapter.

from transfer function models, and vice versa. A discussion of tf2ss and ss2tf appears in Chapter 3 of *Modern Control Systems*.

The system in Eq. (11.1) is stable if the **characteristic roots** of the system all have negative real parts. The characteristic roots are obtained by solving the nth order polynomial equation

$$\det(s\mathbf{I} - \mathbf{A}) = 0 \,. \tag{11.3}$$

The characteristic roots s_i are known as the **eigenvalues** of the matrix $\mathbf{A}$. If the system is defined with the minimal number of state variables, the eigenvalues of $\mathbf{A}$ are exactly the poles of the corresponding transfer function model. Given a system in transfer function form, we determine stability by considering the location of the system poles in the complex plane. Similarly, given a system in state variable form, we determine stability by considering the location in the complex plane of the eigenvalues of the matrix $\mathbf{A}$.

Two other important considerations for state variable models are **controllability** and **observability**. We will discuss these items in Sections 11.3 and 11.4 in more detail. First we present a simple example to illustrate the state variable modeling method.

11.2   ROBOT DRIVE TRAIN DYNAMICS

Due to their compact size and low nonlinear backlash, **harmonic drive trains** are used in robot manipulators [64]. On the negative side, these harmonic drive trains have a significant amount of joint flexibility. An idealized depiction of a drive train is shown in Figure 11.2. The drive train consists of an actuator connected by a torsional spring to a load. The input to the system is the motor torque, $u(t)$. The equations of motion of the idealized system are

$$J_L\ddot{\theta}_L + b_L\dot{\theta}_L + K(\theta_L - \theta_m) = 0 \,, \tag{11.4}$$

$$J_m\ddot{\theta}_m + b_m\dot{\theta}_m + K(\theta_m - \theta_L) = u \,,$$

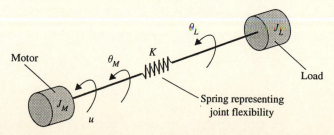

FIGURE 11.2
Depiction of a harmonic drive train.

where b_L and b_m represent the load and motor (torsional or rotational) damping parameters, respectively. The parameters J_L and J_m are the load and motor moments of inertia, respectively. The output of the system is the load angle, θ_L.

To obtain the state variable model, we begin by defining the state variables of the system as

$$x_1 = \theta_L \, ,$$
$$x_2 = \theta_m \, ,$$
$$x_3 = \dot{\theta}_L \, ,$$
$$x_4 = \dot{\theta}_m \, ;$$

the input is u, and the output is θ_L. Note that this assignment of state variables is not unique; we could choose $x_1 = \theta_m$, $x_2 = \theta_L$, and so on.

Taking the time-derivative of x_1 and x_2 yields, respectively,

$$\dot{x}_1 = \dot{\theta}_L = x_3 \tag{11.5}$$

and

$$\dot{x}_2 = \dot{\theta}_m = x_4 \, . \tag{11.6}$$

Similarly, taking the time-derivative of x_3 and x_4 yields

$$\dot{x}_3 = \ddot{\theta}_L = -\frac{b_L}{J_L}\dot{\theta}_L - \frac{K}{J_L}(\theta_L - \theta_m) \tag{11.7}$$

$$= -\frac{b_L}{J_L}x_3 - \frac{K}{J_L}x_1 + \frac{K}{J_L}x_2$$

and

$$\dot{x}_4 = \ddot{\theta}_m = -\frac{b_m}{J_m}\dot{\theta}_m - \frac{K}{J_m}(\theta_m - \theta_L) + \frac{1}{J_m}u \tag{11.8}$$

$$= -\frac{b_m}{J_m}x_4 - \frac{K}{J_m}x_2 + \frac{K}{J_m}x_1 + \frac{1}{J_m}u \, .$$

Combining the results in Eqs. (11.5)–(11.8) yields

$$\dot{\mathbf{x}} = \mathbf{A}\mathbf{x} + \mathbf{B}u \, , \tag{11.9}$$

where

$$
\mathbf{A} = \begin{bmatrix} 0 & 0 & 1 & 0 \\ 0 & 0 & 0 & 1 \\ -\dfrac{K}{J_L} & \dfrac{K}{J_L} & -\dfrac{b_L}{J_L} & 0 \\ \dfrac{K}{J_m} & -\dfrac{K}{J_m} & 0 & -\dfrac{b_m}{J_m} \end{bmatrix} \quad \text{and} \quad \mathbf{B} = \begin{bmatrix} 0 \\ 0 \\ 0 \\ \dfrac{1}{J_m} \end{bmatrix} .
$$

The output equation is

$$
y = \mathbf{C}\mathbf{x} + \mathbf{D}u , \tag{11.10}
$$

where

$$
\mathbf{C} = \begin{bmatrix} 1 & 0 & 0 & 0 \end{bmatrix} \quad \text{and} \quad \mathbf{D} = [0] .
$$

Eqs. (11.9) and (11.10) represent the state variable model. In this case the number of state variables is $n = 4$.

Is the robot drive train system stable? To answer this question, we need to determine the locations of the eigenvalues of the matrix $\mathbf{A}$. Remember that the eigenvalues of $\mathbf{A}$ are identical to the system poles. The characteristic polynomial equation associated with $\mathbf{A}$ is

$$
\det(s\mathbf{I} - \mathbf{A}) = 0 ,
$$

or

$$
s\left[s^3 + \left(\frac{b_L}{J_L} + \frac{b_m}{J_m} \right) s^2 + \left(\frac{b_L b_m}{J_L J_m} + \frac{K}{J_m} + \frac{K}{J_L} \right) s + \frac{K(b_L + b_m)}{J_L J_m} \right] = 0 .
$$

We see immediately that one characteristic root is

$$
s = 0 .
$$

The remaining three characteristic roots are in the left half-plane if

$$
\frac{b_L}{J_L} + \frac{b_m}{J_m} > 0 ,
$$

$$
\left(\frac{b_L}{J_L} + \frac{b_m}{J_m} \right) \frac{b_L b_m}{J_L J_m} + K \left(\frac{b_L}{J_L^2} + \frac{b_m}{J_m^2} \right) > 0 ,
$$

$$
\frac{K(b_L + b_m)}{J_L J_m} > 0 .
$$

These conditions can be derived using Routh-Hurwitz methods (see Chapter 6).

By definition, we have

$$J_L > 0 \ , J_m > 0 \ , \text{ and } K > 0$$

and

$$B_L \geq 0 \text{ and } B_m \geq 0 \ .$$

If either $B_L \neq 0$ or $B_M \neq 0$, three system poles lie in the left half-plane and one pole is at the origin, so the system is stable. If the system has no damping (that is, $B_L = B_m = 0$), then the system characteristic equation reduces to

$$s^2 \left[s^2 + \left(\frac{K}{J_m} + \frac{K}{J_L} \right) \right] = 0 \ .$$

In this case there is a double pole at $s = 0$. The remaining poles are

$$s = \pm j \sqrt{ \frac{K}{J_m} + \frac{K}{J_L} } \ .$$

The double pole at the origin makes the system unstable.

Suppose we have a robot system with the parameters shown in Table 11.1. With the given parameters, the equivalent transfer function model is

$$\frac{\theta_L(s)}{U(s)} = \frac{100}{s^4 + 15s^3 + 29s^2 + 80s} \ .$$

The MATLAB script that we can use to obtain the transfer function model is shown in Figure 11.3. The impulse response is shown in Figure 11.4. At approximately five seconds, the angles $x_1 = \theta_L$ and $x_2 = \theta_M$ settle out at $1.25°$. The angle rates, $x_3 = \dot{\theta}_L$, and $x_4 = \dot{\theta}$ both settle back to zero at steady-state. Using the final value theorem, we verify that θ_L should settle out at $\theta_L = 1.25°$.

TABLE 11.1   Parameters for the Robot Drive Train System

Parameter	Value	Units
K	0.5	ft-lb /deg
J_m	0.1	ft-lb-sec^2 /deg
J_L	0.05	ft-lb-sec^2 /deg
B_m	0.1	ft-lb-sec /deg
B_L	0.7	ft-lb-sec /deg

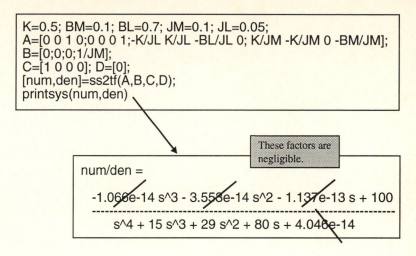

```
K=0.5; BM=0.1; BL=0.7; JM=0.1; JL=0.05;
A=[0 0 1 0;0 0 0 1;-K/JL K/JL -BL/JL 0; K/JM -K/JM 0 -BM/JM];
B=[0;0;0;1/JM];
C=[1 0 0 0]; D=[0];
[num,den]=ss2tf(A,B,C,D);
printsys(num,den)
```

These factors are negligible.

num/den =

-1.066e-14 s^3 - 3.553e-14 s^2 - 1.137e-13 s + 100
----
s^4 + 15 s^3 + 29 s^2 + 80 s + 4.046e-14

FIGURE 11.3
Computing the transfer function model from the state variable model.

The computation is as follows:

$$\lim_{t\to\infty} \theta_L(t) = \lim_{s\to 0} s\theta_L(s) = \lim_{s\to 0} s\frac{100}{s^4 + 15s^3 + 24s^2 + 80s} = \frac{100}{80} = 1.25 \,,$$

where we use the fact that $U(s) = 1$ (that is, $u(t)$ is an impulse). The steady-state value exists because of the pole at the origin. If this nonzero value is unacceptable, then we could use feedback to eliminate it.

11.3   MORE ON CONTROLLABILITY

We say that our system is controllable if, for each initial state $\mathbf{x}(t_o)$ and each final state $\mathbf{x}(t_f)$, there exists an (unconstrained) input $u(t)$ that transfers the system from $\mathbf{x}(t_o)$ at $t = t_o$ to $\mathbf{x}(t_f)$ at $t = t_f$. If a system is controllable, we can move the system from any initial state to any final state in a finite time $t_f - t_o$. Surprisingly, the answer to whether a particular system is controllable is given by a simple algebraic test. A system is controllable if the following rank condition is satisfied:

$$\text{rank}\begin{bmatrix} \mathbf{B} & \mathbf{AB} & \mathbf{A}^2\mathbf{B} & \cdots & \mathbf{A}^{n-1}\mathbf{B} \end{bmatrix} = n \,,$$

where n is the number of state variables. We define the **controllability matrix** $\mathbf{P}_c$ as

$$\mathbf{P}_c = [\ \mathbf{B} \quad \mathbf{AB} \quad \mathbf{A}^2\mathbf{B} \quad \cdots \quad \mathbf{A}^{n-1}\mathbf{B}\ ] \,.$$

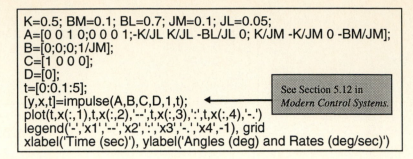

```
K=0.5; BM=0.1; BL=0.7; JM=0.1; JL=0.05;
A=[0 0 1 0;0 0 0 1;-K/JL K/JL -BL/JL 0; K/JM -K/JM 0 -BM/JM];
B=[0;0;0;1/JM];
C=[1 0 0 0];
D=[0];
t=[0:0.1:5];
[y,x,t]=impulse(A,B,C,D,1,t);
plot(t,x(:,1),t,x(:,2),'--',t,x(:,3),':',t,x(:,4),'-.')
legend('-','x1','--','x2',':','x3','-.','x4',-1), grid
xlabel('Time (sec)'), ylabel('Angles (deg) and Rates (deg/sec)')
```

See Section 5.12 in *Modern Control Systems.*

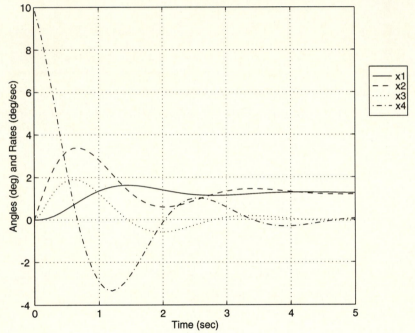

FIGURE 11.4
Impulse response of the harmonic drive train.

For single-input systems, $\mathbf{P}_c$ is a square matrix, and to determine controllability we only need to verify the determinant

$$\det \mathbf{P}_c \neq 0 .$$

How does the notion of controllability affect the design process? A linear state feedback control law has the form

$$u = -\mathbf{H}\mathbf{x} + r , \qquad (11.11)$$

where $\mathbf{H}$ is a gain matrix determined in the design process and r is a reference input. Closing the loop using the control law in Eq. (11.11) yields the closed-loop system

$$\dot{\mathbf{x}} = (\mathbf{A} - \mathbf{BH})\mathbf{x} + \mathbf{B}r \; . \tag{11.12}$$

Since the eigenvalues of $\mathbf{A} - \mathbf{BH}$ are the closed-loop system poles, we would like to be able to locate all the eigenvalues arbitrarily in the left half-plane. In this way we can place the poles of the closed-loop system where they are desired. This design technique is known as **pole placement**.

Suppose we desire the closed-loop poles to have the values

$$-p_1, -p_2, \cdots, -p_n \; ,$$

where p_i may be real or complex. The value of each p_i is selected inside a desired region of the complex plane. The desired region is defined by the performance specifications in the same manner as in the root locus design method. If any specified p_i is a complex number, we must also include the corresponding complex conjugate in the list of desired pole locations.

We write the desired characteristic polynomial as

$$\begin{aligned} p_{\text{des}}(s) &= (s + p_1)(s + p_2) \cdots (s + p_n) \\ &= s^n + \alpha_1 s^{n-1} + \cdots + \alpha_{n-1} s + \alpha_n \; . \end{aligned}$$

There exists a state feedback law (that is, a gain matrix $\mathbf{H}$) such that

$$p_{\text{des}}(s) = \det(s\mathbf{I} - \mathbf{A} + \mathbf{BH}) \; ,$$

if and only if the system is controllable. Obviously, controllability is critical to state feedback design.

We can use **Ackermann's formula** to compute the gain matrix $\mathbf{H}$, which accomplishes the objective of placing the system poles at the desired locations:

$$\mathbf{H} = [\; 0 \quad 0 \quad \cdots \quad 0 \quad 1 \;]\mathbf{P}_c^{-1} p_{\text{des}}(\mathbf{A}) \; , \tag{11.13}$$

where

$$p_{\text{des}}(\mathbf{A}) = \mathbf{A}^n + \alpha_1 \mathbf{A}^{n-1} + \cdots + \alpha_{n-1}\mathbf{A} + \alpha_n \mathbf{I} \; .$$

The number of zeros in the matrix $[0 \quad 0 \quad \cdots \quad 0 \quad 1]$ in Eq. (11.13) is equal to $n - 1$, where n is the number of state variables and $\mathbf{I}$ is an $n \times n$ identity matrix.

If the system is not controllable, then the controllability matrix $\mathbf{P}_c$ will not be invertible (since $\det \mathbf{P}_c = 0$), and the computation of $\mathbf{H}$ using Ackermann's formula (or any other formula) fails.

The MATLAB Control Systems Toolbox has a function, known as the acker function, that calculates the gain matrix $\mathbf{H}$ to place the poles. The acker function is illustrated in Figure 11.5.

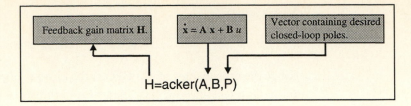

FIGURE 11.5
The acker function.

An illustration of the use of the acker function is shown in Figure 11.6. The desired pole locations are given by

$$p_1 = -10 \,,$$
$$p_2 = -1 \,,$$
$$p_3 = -1 + j \,,$$
$$p_4 = -1 - j \,.$$

The resulting feedback gain matrix is

$$\mathbf{H} = \begin{bmatrix} -3.1 & 3.3 & -4.8 & -0.2 \end{bmatrix} \,.$$

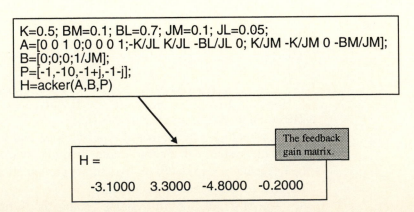

FIGURE 11.6
Using acker to compute **H** to place the poles at $P = (-1, -10, -1 + j, -1 - j)$.

11.4   MORE ON OBSERVABILITY

It is not usually practical (or possible) to sense all the state variables and have them available for feedback. Generally only certain state variables (or linear combinations) are available for feedback. If all the state variables are not available for feedback, the ensuing control design problem is known as an output feedback control design problem. For output feedback to work, we must develop a state observer that provides estimates of the state variables that are not sensed directly. The estimated state variables replace the actual state variables in the state feedback control law. The subject of output feedback control is a very rich and interesting field of study. We will not, however, cover the subject here. One of the original publications on the subject of observers is by Luenberger [65]. A good overview can be found in Levine [66].

A system is observable if and only if there exists a finite time T such that the initial state $\mathbf{x}(t_o)$ can be determined from the observation of $y(t)$, given the control $u(t)$. As with controllability, we can determine the observability of a system by a simple algebraic test. Define the **observability matrix Q** as

$$
\mathbf{Q} = \begin{bmatrix} \mathbf{C} \\ \mathbf{CA} \\ \mathbf{CA}^2 \\ \vdots \\ \mathbf{CA}^{n-1} \end{bmatrix} .
$$

For single-output systems, $\mathbf{Q}$ is an $n \times n$ matrix. A system is said to be observable when

$$
\text{rank } \mathbf{Q} = n .
$$

For single-output systems, the observability condition becomes $\det \mathbf{Q} \neq 0$.

What can we say about a system that is not controllable and observable? Consider the system

$$
\dot{\mathbf{x}} = \mathbf{Ax} + \mathbf{B}u ,
$$
$$
y = \mathbf{Cx} + \mathbf{D}u ,
$$

where

$$
\mathbf{A} = \begin{bmatrix} 2 & 0 \\ 1 & 1 \end{bmatrix} , \quad \mathbf{B} = \begin{bmatrix} 0 \\ 1 \end{bmatrix} , \quad \mathbf{C} = [\ 1 \quad -1\ ] , \text{ and } \mathbf{D} = [0] .
$$

The system is neither controllable nor observable. Use the MATLAB functions ctrb and obsv to determine the controllability and observability, respectively (see Chapter 11 in *Modern Control Systems*).

The transfer function associated with the state variable model is the first-order transfer function given by

$$G(s) = \mathbf{C}(s\mathbf{I} - \mathbf{A})^{-1}\mathbf{B}$$
$$= \frac{-1}{s - 1}.$$

The state variable model has two state variables (that is, $n = 2$), but the transfer function model is a first-order model. This shows that one of the state variables in the state variable model is redundant—it provides no additional information. The system that is not observable and controllable is not of minimal dimension.

A minimal dimension state variable model can be derived from $G(s)$. One possibility is

$$\dot{\mathbf{x}} = \mathbf{A}\mathbf{x} + \mathbf{B}u\,,$$
$$y = \mathbf{C}\mathbf{x} + \mathbf{D}u\,,$$

where

$$\mathbf{A} = [1]\,, \quad \mathbf{B} = [1]\,,$$
$$\mathbf{C} = [-1]\,, \quad \mathbf{D} = [0]\,.$$

This system is controllable, observable, and has only one state variable ($n = 1$).

11.5 DIESEL ELECTRIC LOCOMOTIVE EXAMPLE

The diesel electric locomotive example is adapted from DP11.3 in *Modern Control Systems*. The system is depicted in Figure 11.7. The efficiency of the diesel engine is very sensitive to the speed of rotation of the motors. We want to design a control system that drives the electric motors of a diesel electric locomotive for use on railroad trains. The locomotive is driven by dc motors located on each of the axles. The throttle position (see Figure 11.7) is set by moving the input potentiometers.

The control objective is to regulate the shaft rotation speed, denoted by ω_o, to the desired value, ω_r.

Control Goal   Regulate the shaft rotation speed to the desired value in the presence of external load torque disturbances.

The corresponding variable to be controlled is the shaft rotation speed, or ω_o.

Variable to Be Controlled

Shaft rotation speed, ω_o.

The controlled speed, ω_o, is sensed by a tachometer, which supplies a feedback voltage v_o. The electronic amplifier amplifies the error signal, $v_r - v_o$, between the reference and feedback voltage signals and provides a voltage v_f that is supplied to the field winding of a dc generator.

The generator is run at a constant speed ω_d by the diesel engine and generates a voltage v_g that is supplied to the armature of a dc motor. The motor is armature controlled, with a fixed current supplied to its field. As a result, the motor produces a torque T and drives the load connected to its shaft so that the controlled speed, ω_o, tends to equal the command speed, ω_r.

A signal flow diagram of the sytem is shown in Figure 11.8. In Figure 11.8 we use L_t and R_t, which are defined as

$$L_t = L_a + L_g \, ,$$
$$R_t = R_a + R_g \, .$$

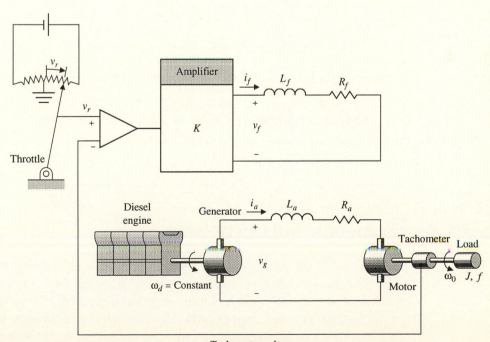

v_o = Tachometer voltage

FIGURE 11.7
Diesel electric locomotive system.

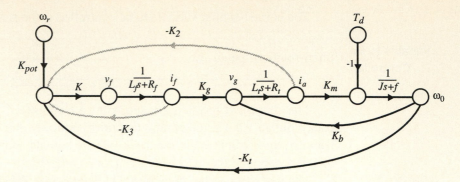

FIGURE 11.8
Signal flow graph of the diesel electric locomotive.

Values for the parameters of the diesel electric locomotive are given in Table 11.2.

Notice that the system has a feedback loop; we use the tachometer voltage, v_o, as a feedback signal to form an error signal, $v_r - v_o$. Without additional state feedback, the only tuning parameter is the amplifier gain K. As a first step, we can investigate the system performance with tachometer voltage feedback only.

The key tuning parameters are given by

Select Key Tuning Parameters	K and $\mathbf{H}$

The matrix $\mathbf{H}$ is the state feedback gain matrix. The design specifications are

Design Specifications	**DS1** Steady-state tracking error less than 2% to a unit step input.
	DS2 Percent overshoot of $\omega_o(t)$ less than 10% to a unit step input $\omega_r(s) = 1/s$.
	DS3 Settling time less than 1 second to a unit step input.

11.6 STATE FEEDBACK CONTROLLER DESIGN

The first step in the development of the vector differential equation that accurately describes the system is to choose a set of state variables. In practice the selection of state variables can be a difficult process, especially for complex systems. The state variables must be sufficient in number to determine the future behavior of the system when the present state and all future inputs are known. The selection of state variables is intimately related to the issue of complexity. How complex should a model be for the control system design? This is not a question that can be adequately addressed here, but it is an important topic.

TABLE 11.2   Parameter Values for the Diesel Electric Locomotive.

K_m	K_g	K_b	J	f	L_a	R_a	R_f	L_f	K_t	K_{pot}	L_g	R_g
10	100	0.62	1	1	0.2	1	1	0.1	1	1	0.1	1

The diesel electric locomotive system has three major components: two electrical circuits and one mechanical system. It seems logical that the state vector will include state variables from both electrical circuits and from the mechanical system. One reasonable choice of state variables is

$$x_1 = \omega_o \,,$$
$$x_2 = i_a \,,$$
$$x_3 = i_f \,.$$

This state variable selection is not unique. With the state variables defined above, the state variable model is

$$\dot{x}_1 = -\frac{f}{J}x_1 + \frac{K_m}{J}x_2 - \frac{1}{J}T_d \,,$$
$$\dot{x}_2 = -\frac{K_b}{L_t}x_1 - \frac{R_t}{L_t}x_2 + \frac{K_g}{L_t}x_3 \,,$$
$$\dot{x}_3 = -\frac{R_f}{L_f}x_3 + \frac{1}{L_f}u \,,$$

where

$$u = K K_{pot}\omega_r - K K_t x_1 \,.$$

In matrix form (with $T_d(s) = 0$), we have

$$\dot{\mathbf{x}} = \mathbf{A}\mathbf{x} + \mathbf{B}u \,, \tag{11.14}$$
$$y = \mathbf{C}\mathbf{x} + \mathbf{D}u \,,$$

where

$$\mathbf{A} = \begin{bmatrix} -\frac{f}{J} & \frac{K_m}{J} & 0 \\ -\frac{K_b}{L_t} & -\frac{K_t}{L_t} & \frac{K_g}{L_t} \\ 0 & 0 & -\frac{R_f}{L_f} \end{bmatrix} \,, \quad \mathbf{B} = \begin{bmatrix} 0 \\ 0 \\ \frac{1}{L_f} \end{bmatrix} \,, \text{ and}$$

$$\mathbf{C} = [\, 1 \quad 0 \quad 0 \,], \quad \mathbf{D} = [0] \,.$$

The corresponding transfer function is

$$G(s) = \mathbf{C}(s\mathbf{I} - \mathbf{A})^{-1}\mathbf{B} = \frac{K_g K_m}{(R_f + L_f s)[(R_t + L_t s)(Js + f) + K_m K_b]} .$$

If we take advantage of the fact that

$$K_{\text{pot}} = K_t = 1 ,$$

then (from an input-output perspective) the system has the simple feedback configuration shown in Figure 11.9.

Using the parameter values given in Table 11.2 and computing the steady-state tracking error for a unit step input yields

$$e_{ss} = \frac{1}{1 + KG(0)} = \frac{1}{1 + 121.95K} .$$

Using the Routh-Hurwith method, we also find that the closed-loop system is stable for

$$-0.008 < K < 0.0468 .$$

The smallest steady-state tracking error is achieved for the largest value of K. At best we can obtain a 15% tracking error, which does not meet the design specification DS1. Also, as K gets larger, the response becomes unacceptably oscillatory.

We now consider a **state feedback controller design**. The feedback loops are shown in Figure 11.8, which shows that ω_o, i_a, and i_f are available for feedback. Without any loss of generality, we set $K = 1$. Any value of $K > 0$ would work as well.

The control input is

$$u = K_{pot}\omega_r - K_t x_1 - K_2 x_2 - K_3 x_3 .$$

The feedback gains to be determined are K_t, K_2, and K_3. The tachometer gain, K_t, is now a key parameter of the design process. Also K_{pot} is a key variable for

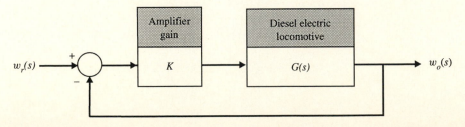

FIGURE 11.9
Block diagram representation of the diesel electric locomotive.

tuning. By adjusting the parameter K_{pot}, we have the freedom to scale the input ω_r. When we define

$$\mathbf{H} = [\ K_t \quad K_2 \quad K_3\ ],$$

then

$$u = -\mathbf{H}\mathbf{x} + K_{pot}\omega_r\ . \tag{11.15}$$

The closed-loop system with state feedback is

$$\dot{\mathbf{x}} = (\mathbf{A} - \mathbf{B}\mathbf{H})\mathbf{x} + \mathbf{B}v\ ,$$
$$y = \mathbf{C}\mathbf{x}\ ,$$

where

$$v = K_{pot}\omega_r\ .$$

We will use pole-placement methods to determine $\mathbf{H}$ such that the eigenvalues of $\mathbf{A} - \mathbf{B}\mathbf{H}$ are in the desired locations. First we make sure the system is controllable. When $n = 3$ the controllability matrix is

$$\mathbf{P}_c = [\ \mathbf{B} \quad \mathbf{A}\mathbf{B} \quad \mathbf{A}^2\mathbf{B}\ ]\ .$$

Computing the determinant of P_c yields

$$\det \mathbf{P}_c = -\frac{K_g^2 K_m}{JL_f^3 L_t^2}\ .$$

Since $K_g \neq 0$ and $K_m \neq 0$ and $JL_f^3 L_t^2$ is nonzero, we determine that

$$\det \mathbf{P}_c \neq 0\ .$$

Thus the system is controllable. We can place all the poles of the system appropriately to satisfy DS2 and DS3.

The desired region to place the eigenvalues of $\mathbf{A} - \mathbf{B}\mathbf{H}$ is illustrated in Figure 11.10. The specific pole locations are selected to be

$$p_1 = -50\ ,$$
$$p_2 = -4 + 3j\ ,$$
$$p_3 = -4 - 3j\ .$$

Selecting $p_1 = 50$ allows for a good second-order response that is governed by p_2 and p_3.

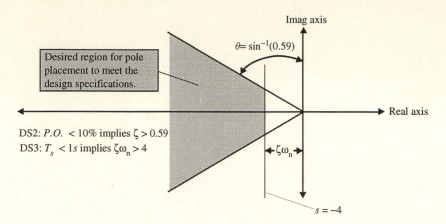

FIGURE 11.10
Desired location of the closed-loop poles (that is, the eigenvalues of $\mathbf{A} - \mathbf{BH}$).

The gain matrix $\mathbf{H}$ that achieves the desired closed-loop poles is

$$H = \begin{bmatrix} -0.0041 & 0.0035 & 4.0333 \end{bmatrix} .$$

To select the gain K_{pot}, we first compute the dc gain of the closed-loop transfer function. With the state feedback in place, the closed-loop transfer function is

$$T(s) = \mathbf{C}(s\mathbf{I} - \mathbf{A} + \mathbf{BH})^{-1}\mathbf{B} .$$

Then

$$K_{pot} = \frac{1}{T(0)} .$$

Using the gain K_{pot} in this manner effectively scales the closed-loop transfer function so that the dc gain is equal to 1. We then expect that a unit step input representing a 1°/sec step command results in a 1°/sec steady-state output at ω_o.

The step response of the system is shown in Figure 11.11. We can see that all the design specifications are satisfied.

It would be an interesting exercise to conduct numerical experiments to study the effect of varying the gain matrix $\mathbf{H}$. This can certainly be done with the script given in Figure 11.11. The pole locations can be varied easily within the MATLAB script and the associated gain matrix automatically computed using the acker function. As we see in the next section, it is also possible to conduct numerical experiments using a simulation of the system developed in SIMULINK.

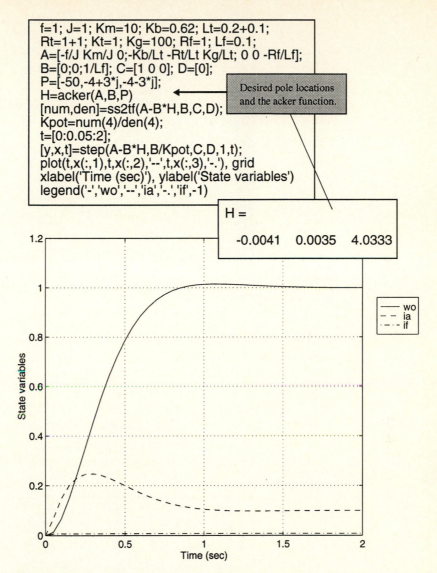

```
f=1; J=1; Km=10; Kb=0.62; Lt=0.2+0.1;
Rt=1+1; Kt=1; Kg=100; Rf=1; Lf=0.1;
A=[-f/J Km/J 0;-Kb/Lt -Rt/Lt Kg/Lt; 0 0 -Rf/Lf];
B=[0;0;1/Lf]; C=[1 0 0]; D=[0];
P=[-50,-4+3*j,-4-3*j];
H=acker(A,B,P)
[num,den]=ss2tf(A-B*H,B,C,D);
Kpot=num(4)/den(4);
t=[0:0.05:2];
[y,x,t]=step(A-B*H,B/Kpot,C,D,1,t);
plot(t,x(:,1),t,x(:,2),'--',t,x(:,3),'-.'), grid
xlabel('Time (sec)'), ylabel('State variables')
legend('-','wo','--','ia','-.','if',-1)
```

Desired pole locations
and the acker function.

H =

    -0.0041   0.0035   4.0333

FIGURE 11.11
Closed-loop step response of the diesel electric locomotive.

11.7   SYSTEM SIMULATION WITH SIMULINK

The SIMULINK block diagram of the closed-loop system is shown in Figure 11.12. The system model is given in terms of the state variable representation (that is, **A** and **B**) . The outputs of the plant block are the three states ω_o, i_a, and i_f. These states are fed back to the feedback gain matrix block to obtain the feedback sig-

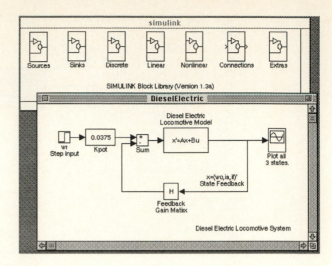

FIGURE 11.12
SIMULINK basic block diagram.

nal **Hx**. The input signal ω_r is scaled by K_{pot} and summed with the feedback signal forming the plant input signal.

We can adjust the value of the gain K_{pot} with a slider control, as shown in Figure 11.13, in which $K_{pot} = 0.0375$. The slider control makes the adjustment of the parameter a simple process. The gain matrix **H** and the system matrices **A** and **B** are set by selecting the appropriate block and inputing the data into the spaces provided. This is shown in Figures 11.14 and 11.15.

Once we have defined all the parameters, we can initiate the simulation (see

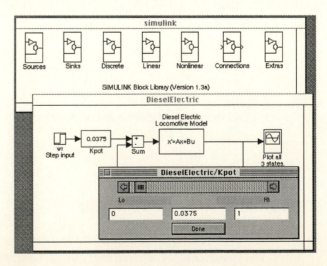

FIGURE 11.13
Modifying the value of K_{pot} using a slider control.

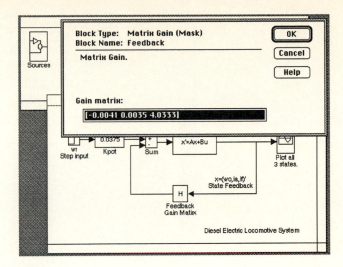

FIGURE 11.14
Incorporating the feedback gain matrix into the simulation block diagram.

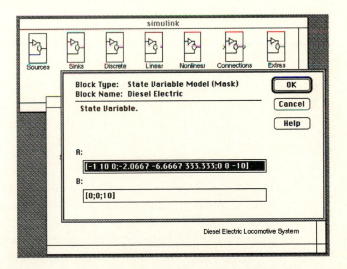

FIGURE 11.15
Defining the state variable model via the matrices **A** and **B**.

Chapters 1 and 5 for more information on starting a simulation and setting the simulation parameters). The resulting state variable time histories are shown in Figure 11.16. The plot is the same as we obtained using the MATLAB script given in Figure 11.11. We can use the SIMULINK simulation to conduct numerical experiments easily and to analyze the closed-loop system performance under a variety of conditions.

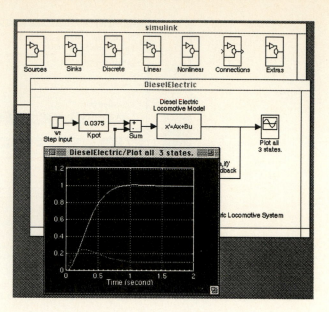

FIGURE 11.16
The result of the simulation showing the state variable history (the graph corresponds to Figure 11.11).

11.8   SUMMARY

In this chapter we discussed state variable models. The system poles are the same as the eigenvalues of the system matrix **A** when the state variable model is of minimal dimension. We used a simple example based on robot drive train dynamics to illustrate the state variable modeling process. A review of controllability and observability stressed their importance in state feedback control system design.

A necessary and sufficient condition for arbitrarily placing the closed-loop poles using state feedback is that the system be controllable. Given a controllable system, we can use Ackermann's formula to obtain the feedback gain matrix that places the closed-loop poles at the desired locations. The chapter ended with the design of a state feedback controller for a diesel electric locomotive problem. Using the acker function, we can easily compute the feedback gain matrix. We presented a SIMULINK simulation to assist in the analysis of the control design and to provide a mechanism to perform numerical experiments and answer what if questions.

E X E R C I S E S

E11.1 Consider a system represented in state variable form:

$$\dot{x} = \mathbf{A}x + \mathbf{B}u ,$$
$$y = \mathbf{C}x + \mathbf{D}u .$$

Determine if the following systems are controllable and observable:

(a)

$$\mathbf{A} = \begin{bmatrix} 0 & 1 & 0 & 0 \\ 0 & 0 & 1 & 0 \\ 0 & 0 & 0 & 1 \\ -1 & -4 & 0 & -3 \end{bmatrix} , \quad \mathbf{B} = \begin{bmatrix} 0 \\ 0 \\ 0 \\ 10 \end{bmatrix} ,$$

$$\mathbf{C} = [\, 1 \;\; 0 \;\; 0 \;\; 0 \,] , \quad \mathbf{D} = [0] .$$

(b)

$$\mathbf{A} = \begin{bmatrix} 0 & 1 \\ 1 & 1.5 \end{bmatrix} , \quad \mathbf{B} = \begin{bmatrix} 1 \\ 2 \end{bmatrix} ,$$

$$\mathbf{C} = [\, 1 \;\; 0 \,] , \quad \mathbf{D} = [0] .$$

E11.2 Consider the system

$$\dot{x} = \begin{bmatrix} 1 & 2 \\ -2 & 4 \end{bmatrix} x + \begin{bmatrix} 1 \\ q \end{bmatrix} u .$$

For what values of q is the system controllable?

E11.3 Compute the state feedback gain matrix that places the poles of the system

$$\dot{x} = \begin{bmatrix} 0 & 1 \\ -2 & -3 \end{bmatrix} x + \begin{bmatrix} 1 \\ 1 \end{bmatrix} u$$

at $p_1 = -1$ and $p_2 = -10$.

E11.4 Consider the diesel electric locomotive problem discussed in Section 11.5. Design a state feedback controller so the system step response is overdamped. The system must still settle out in under 1 second. Use MATLAB or SIMULINK to analyze the closed-loop system performance with the controller in the loop.

CHAPTER 12

Robust Control Systems

Digital Audio Tape Speed Control Example

12.1 Introduction 216
12.2 Uncertain Time-delays 217
12.3 Digital Audio Tape Example 222
12.4 PID Controller Design 224
12.5 Summary 229
    Exercises 229

P R E V I E W

In this chapter we use the digital audio tape speed control problem to study the robustness of a PID controller to changes in several important plant parameters. The PID controller design is accomplished using root locus methods on the nominal plant. Also, the idea of robust stability is discussed in the context of uncertain time-delay models. For testing robust stability with uncertain time-delays we present a method based on the so-called small gain theorem.

12.1 INTRODUCTION

The issue of **robustness** is critical to the ultimate success of the control system. The control system may be required to perform in situations for which it was not originally intended. The fact is, mathematical models of physical systems have limitations. The models are based on a set of assumptions that may or may not

216

reflect reality. The difference between the physical system and the mathematical model arises from many sources, including

- Parameter changes
- Unmodeled dynamics
- Unmodeled sensor noise and biases
- Unexpected disturbances
- Unmodeled time-delays
- Changes in the equilibrium point

We want our closed-loop system to be stable and to retain acceptable performance in the presence of **model uncertainties**. We use the phrase "stability robustness" when considering the system stability in the presence of uncertainty and "performance robustness" when considering the system performance in the presence of uncertainty. A significant amount of progress has been made in the design and analysis of robust control systems since the 1980s [67].

The essential elements of the design process emphasized in this chapter are shown in Figure 12.1. The issue of designing robust control systems is a very interesting and advanced subject. Some popular design methods are known (in the controls community) as LQG/LTR, H_∞, and L^1. These names may seem a bit obscure to the uninitiated, but suffice it to say that they represent practical design methodologies that are becoming more commonplace in industry. We will restrict our attention to robust stability analysis: Given a controller, is the system stable for a given set of uncertainties? One important class of uncertainties is due to time-delays, e^{-Ts}, where T is known to lie in a range $T_1 \leq T \leq T_2$. Uncertain time-delays are the topic of the next section.

12.2   UNCERTAIN TIME-DELAYS

Consider the feedback control system shown in Figure 12.2, where

$$G_d(s) = e^{-Ts} .$$

The exact value of the time-delay is uncertain, but it is known to lie in the interval $T_1 \leq T \leq T_2$. For example, if a robot on Mars is being remotely controlled from earth, the time it takes the signals to reach the planetary robot is not precisely known since transient time depends on the distance between the transmitter and the planetary robot, the atmospheric medium through which the signals travel, interplanetary space effects, and so forth—all of which are time varying and cannot be precisely modeled.

In the development of a robust stability controller, we would like to represent the **time-delay uncertainty** in the form shown in Figure 12.3, where we need to determine a function $M(s)$ that approximately models the time-delay. This will

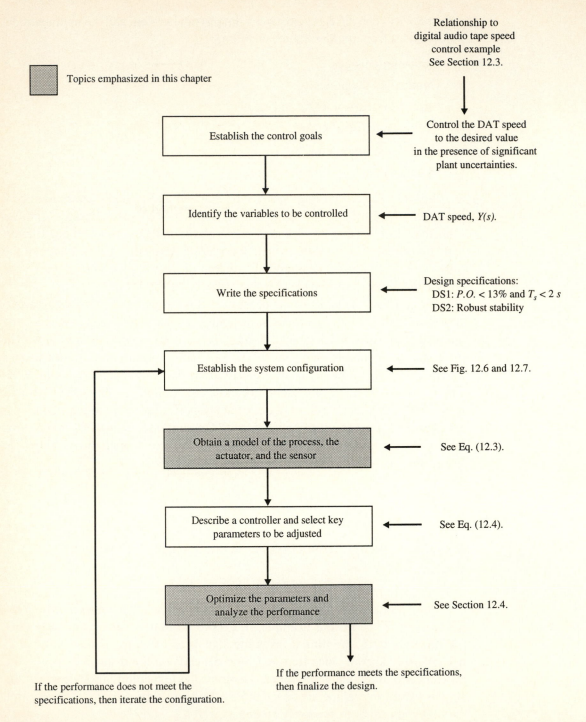

FIGURE 12.1
Elements of the control system design process emphasized in this chapter.

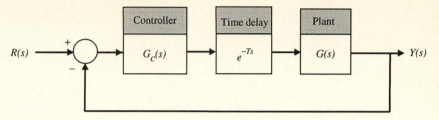

FIGURE 12.2
A feedback system with a time-delay in the loop.

lead to the establishment of a straightforward method of testing the system for stability robustness in the presence of the uncertain time-delay. The uncertainty model is known as a **multiplicative uncertainty representation**.

Since we are concerned with stability, we can consider $R(s) = 0$. Then we can manipulate the block diagram in Figure 12.3 to obtain the form shown in Figure 12.4. Using the so-called **small gain theorem**, we have the condition that the closed-loop system is stable if

$$|M(j\omega)| \; |\frac{G_c(j\omega)G(j\omega)}{1 + G_c(j\omega)G(j\omega)}| \; < \; 1 ,$$

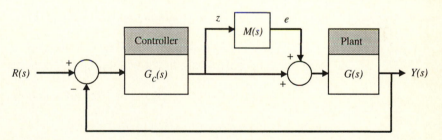

FIGURE 12.3
Multiplicative uncertainty representation.

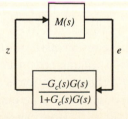

FIGURE 12.4
Equivalent block diagram depiction of the multiplicative uncertainty.

or equivalently (see Chapter 12 in *Modern Control Systems*)

$$|M(j\omega)| < \left|1 + \frac{1}{G_c(j\omega)G(j\omega)}\right| \quad \text{for all } \omega .$$

The question is how to apply the small gain theorem to the analysis of the stability of the system with the uncertain time-delay. Define

$$G_m(s) = e^{-Ts}G(s) .$$

Then

$$G_m(s) - G(s) = e^{-Ts}G(s) - G(s) = (e^{-Ts} - 1)G(s) ,$$

or

$$\frac{G_m(s)}{G(s)} - 1 = e^{-Ts} - 1 .$$

If we define

$$M(s) = e^{-Ts} - 1 ,$$

then we have

$$G_m(s) = (1 + M(s))G(s) . \tag{12.1}$$

The problem is that the time-delay T is not known exactly. One approach to solving the problem is to find a weighting function, denoted by $W(s)$, such that

$$|e^{-j\omega T} - 1| < |W(j\omega)| \quad \text{for all } \omega \text{ and } T_1 \leq T \leq T_2 .$$

If $W(s)$ satisfies the above inequality, we have

$$|M(j\omega)| < |W(j\omega)| .$$

Therefore, the robust stability condition is given by

$$|W(j\omega)| < \left|1 + \frac{1}{G_c(j\omega)G(j\omega)}\right| \quad \text{for all } \omega . \tag{12.2}$$

This is a **conservative bound**. If the condition in Eq. (12.2) is satisfied, then stability is guaranteed in the presence of any time-delay in the range $T_1 \leq T \leq T_2$. If the condition is not satisfied, the system may or may not be stable.

Suppose we have an uncertain time-delay that is known to lie in the range $0.1 \leq T \leq 1$. How can we determine a suitable weighting function $W(s)$? A plot of the magnitude of $e^{-j\omega T} - 1$ is shown in Figure 12.5 for various values of T in the range $T_1 \leq T \leq T_2$. A reasonable weighting function obtained by trial and error is

$$W(s) = \frac{2.5s}{1.2s + 1} \, .$$

This function satisfies the condition

$$|e^{-J\omega T} - 1| \; < \; |W(J\omega)| \, .$$

Keep in mind that the selection of the weighting function is not unique.

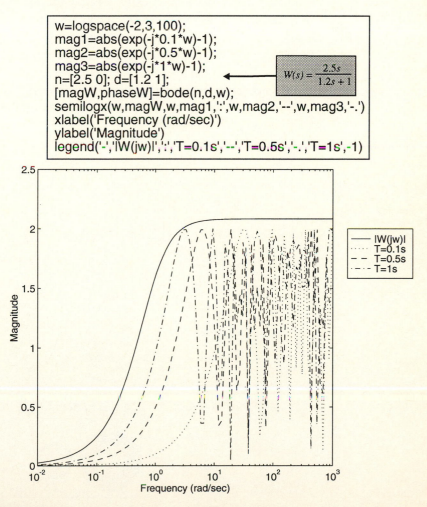

FIGURE 12.5
Magnitude plot of $e^{-J\omega T} - 1$ for $T = 0.1, 0.5,$ and 1.

12.3 DIGITAL AUDIO TAPE EXAMPLE

The digital audio tape problem is adapted from DP12.2 in *Modern Control Systems*. A digital audio tape (DAT) stores 1.3 gigabytes of data in a package the size of a credit card—roughly nine times more than a half-inch-wide reel-to-reel tape or quarter-inch-wide cartridge tape. A DAT sells for about the same amount as a floppy disk, even though it can store 1000 times more data. A DAT can record for two hours (longer than either reel-to-reel or cartridge tape), which means that it can run longer unattended and requires fewer changes and hence fewer interruptions of data transfer. DAT gives access to a given data file within 20 seconds, on the average, compared with several minutes for either cartridge or reel-to-reel tape [68].

The tape drive electronically controls the relative speeds of the drum and tape so that the heads follow the tracks on the tape, as shown in Figure 12.6. The control system is much more complex than that for a CD-ROM because more motors have to be accurately controlled: capstan, take-up and supply reels, drum, and tension control.

Consider the speed control system shown in Figure 12.7. The motor and load transfer function varies because the tape moves from one reel to the other. The transfer function is

$$G(s) = \frac{K_m}{(s + p_1)(s + p_2)}, \tag{12.3}$$

where nominal values are

$$K_m = 4, \quad p_1 = 1, \quad p_2 = 4.$$

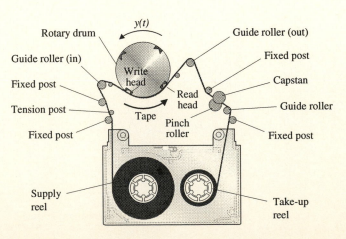

FIGURE 12.6
Digital audio tape driver mechanism.

However, the **range of variation** is

$$3 \le K_m \le 5 \,,$$
$$0.5 \le p_1 \le 1.5 \,,$$
$$3.5 \le p_2 \le 4.5 \,.$$

Thus, the plant is actually a family of plants where each member corresponds to different values of K_m, p_1, and p_2. The design goal is

Design Goal   Control the DAT speed to the desired value in the presence of significant plant uncertainties.

Associated with the design goal we have the variable to be controlled defined as the tape speed:

Variable to Be Controlled   DAT speed, $Y(s)$.

The design specifications are

Design Specifications   **DS1** Percent overshoot less than 13% and settling time less than 2 seconds for a unit step input.

DS2 Robust stability in the presence of a time-delay at the plant input. The time-delay value is uncertain but known to be in the range $0 \le T \le 0.1$.

Design specification DS1 must be satisfied for all plants in the family. Design specification DS2 must be satisfied by the nominal plant ($K_m = 4$, $p_1 = 1$, $p_2 = 4$).

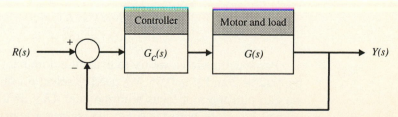

FIGURE 12.7
Block diagram of the digital audio tape speed control system.

The following constraints on the design are given:

- Fast peak time requires that an overdamped condition is not acceptable.
- Use a PID controller:

$$G_c(s) = K_P + \frac{K_I}{s} + K_D s \,. \tag{12.4}$$

- $K_m K_D \leq 20$ when $K_m = 4$.

The key tuning parameters are the PID gains:

Select Key Tuning Parameters K_P, K_I, and K_D.

12.4 PID CONTROLLER DESIGN

Since we are constrained to have $K_m K_D \leq 20$ when $K_m = 4$, we must select $K_D \leq 5$. We will design the PID controller using nominal values for K_m, p_1, and p_2. We will analyze the performance of the controlled system for the various values of the plant parameters, using a simulation to check that DS1 is satisfied. The nominal plant is given by

$$G(s) = \frac{4}{(s + 1)(s + 4)} \,.$$

The closed-loop transfer function is

$$T(s) = \frac{4K_D s^2 + 4K_P s + 4K_I}{s^3 + (5 + 4K_D)s^2 + (4 + 4K_P)s + 4K_I} \,.$$

If we choose $K_D = 5$, then we write the characteristic equation as

$$s^3 + 25s^2 + 4s + 4(K_P s + K_I) = 0 \,,$$

or

$$1 + \frac{4K_P(s + K_I/K_P)}{s(s^2 + 25s + 4)} = 0 \,.$$

Per specifications, we try to place the dominant poles in the region defined by $\zeta \omega_n > 2$ and $\zeta > 0.55$. We need to select a value of $\tau = K_I/K_P$, and then we can plot the root locus with the gain $4K_P$ as the varying parameter. After several iterations with MATLAB, we choose a reasonable value of $\tau = 3$. The root locus is shown in Figure 12.8. Using the rlocfind function, we determine that $4K_P \approx 120$ represents a valid selection since the roots lie inside the desired

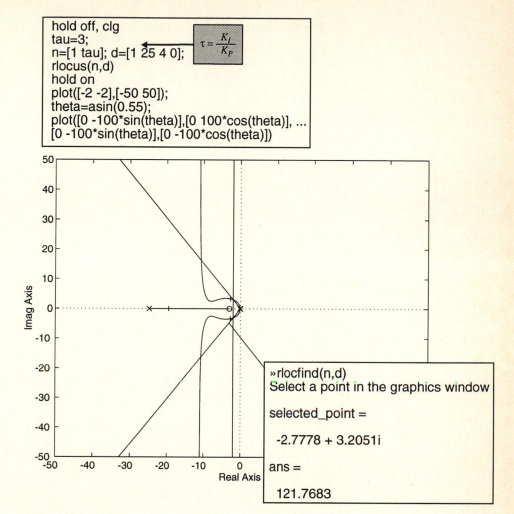

FIGURE 12.8
Root locus for the DAT system with $K_D = 5$ and $\tau = K_I / K_P = 3$.

performance region. This value of $4K_P$ has been rounded off from the exact value on the root locus plot of $4K_P = 121.7683$. We obtain

$$K_P = 30 \, ,$$

and

$$K_I = \tau K_P = 90 \, .$$

The PID controller is then given by

$$G_c(s) = 30 + \frac{90}{s} + 5s \, . \tag{12.5}$$

The step response (for the plant with nominal parameter values) is shown in Figure 12.9. A family of responses is shown in Figure 12.10 for various values of K_m, p_1, and p_2. None of the responses suggests a percent overshoot over the specified value of 13%, and the settling times are all under the 2 second specification as well. As we can see in Figure 12.10, all of the tested plants in the family are adequately controlled by the single PID controller in Eq. (12.5). Therefore DS1 is satisfied for all plants in the family.

Suppose the system has a time-delay at the input to the plant. The actual time-delay is uncertain but known to be in the range $0 \leq T \leq 0.1$ seconds. Following the method discussed in Section 12.2, we determine that a reasonable

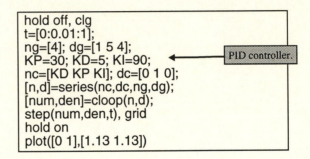

```
hold off, clg
t=[0:0.01:1];
ng=[4]; dg=[1 5 4];
KP=30; KD=5; KI=90;          ← PID controller.
nc=[KD KP KI]; dc=[0 1 0];
[n,d]=series(nc,dc,ng,dg);
[num,den]=cloop(n,d);
step(num,den,t), grid
hold on
plot([0 1],[1.13 1.13])
```

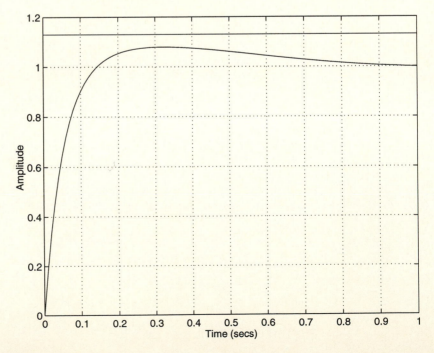

FIGURE 12.9
Unit step response for the DAT system with $K_P = 30$, $K_D = 5$, and $K_I = 90$.

```
hold off, clg
KP=30; KD=5; KI=90;
Km=[3:0.5:5];
P1=[0.5:0.5:1.5];        ⎫    Varying plant parameters:
P2=[3.5:0.5:4.5];        ⎬    Km, p1, and p2.
N=length(Km)*length(P1)*length(P2);   ⎭
t=[0:0.01:2];
nc=[KD KP KI]; dc=[0 1 0];
hold on
for i=1:length(Km)
for j=1:length(P1)
for k=1:length(P2)
  ng=[Km(i)]; dg=[1 P1(j)+P2(k) P1(j)*P2(k)];
  [n,d]=series(nc,dc,ng,dg);[num,den]=cloop(n,d);
  step(num,den,t)
end; end; end
plot([0 2],[1.13 1.13]), grid
```

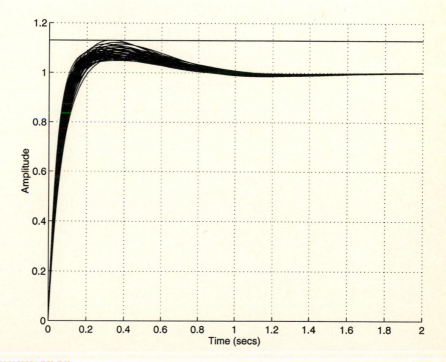

FIGURE 12.10

A family of step responses for the DAT system for various values of the plant parameters K_m, p_1, and p_2.

function $W(s)$ which bounds the plots of $|e^{-j\omega T} - 1|$ for various values of T is

$$W(s) = \frac{0.29s}{0.28s + 1}.$$

To check the stability robustness property, we need to verify that

$$|W(j\omega)| < \left|1 + \frac{1}{G_c(j\omega)G(j\omega)}\right| \quad \text{for all } \omega. \tag{12.6}$$

The plot of both $|W(j\omega)|$ and $\left|1 + \frac{1}{G_c(j\omega)G(j\omega)}\right|$ is shown in Figure 12.11.

It can be seen that the condition in Eq. (12.6) is indeed satisfied. Therefore, we expect that the nominal system will remain stable in the presence of time-delays up to 0.1 seconds.

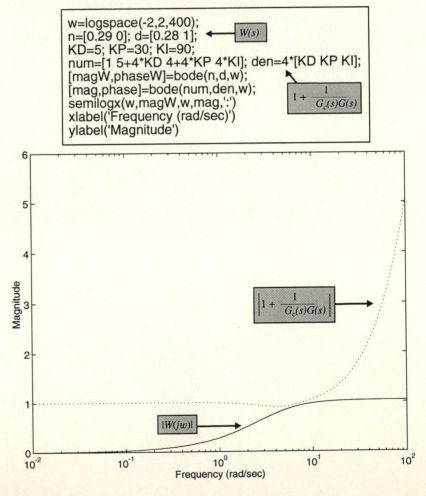

FIGURE 12.11
Stability robustness to a time-delay of uncertain magnitude.

12.5   SUMMARY

In this chapter we investigated the robustness of a controlled system to changes in plant parameter values. The digital audio tape speed control problem was used in the study. We also took a deeper look into stability robustness in the presence of uncertain time-delays. Using a multiplicative perturbation model, we showed that the problem could be formulated in such a way that a relatively simple check can be used to obtain guarantees of stability over the given range of possible time-delays.

E X E R C I S E S

E12.1   Obtain a weighting function $W(s)$ that can be used in stability robustness analysis to represent an uncertain time-delay $0.1 \leq T \leq 0.5$ seconds (see Section 12.2).

E12.2   For the system in Figure E12.2, derive a sufficient condition for stability in the presence of the time-delay.

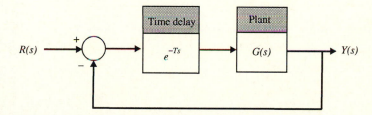

FIGURE E12.2
Block diagram of the simple feedback system with a time-delay.

CHAPTER 13

Digital Control Systems

Fly-by-wire Control Surface Example

13.1 Introduction   230

13.2 Fly-by-wire Aircraft Control Surface   231

13.3 Settling Time and Percent Overshoot Specifications   236

13.4 Controller Design   237

13.5 Summary   240

Exercises   241

PREVIEW

In this chapter we use the fly-by-wire aircraft control surface problem to illustrate the root locus design method for digital control system design. Similar to the s-plane design approach, the z-plane root locus approach places the dominant poles in the desired z-plane region defined by the performance specifications. We discuss the procedure for determining the desired regions in the z-plane to meet settling time and percent overshoot specifications.

13.1   INTRODUCTION

In many control systems the controller is implemented in a digital computer. The computer is the compensator. Computer-controlled systems are becoming very common. The Boeing 777 commercial transport is a fully integrated fly-by-wire airplane (although there are analog back-up systems using old-fashioned cables and pulleys). Most new automobiles have computers on board performing control functions such as ignition, fuel injection, ride control, and so on. Obviously the study of digital control systems is an important area for control engineers. Chapter 13 in *Modern Control Systems* is meant as a brief introduction to digital control systems. We attempt in this chapter to add to the discussion by presenting the design of a digital control system for a fly-by-wire aircraft control surface. But keep in mind that digital control system design is a rich field of study with many real-world applications and as such deserves a significant amount of independent study.

The elements of the design process emphasized in the chapter are shown in Figure 13.1.

13.2   FLY-BY-WIRE AIRCRAFT CONTROL SURFACE

Increasing constraints on weight, performance, fuel consumption, and reliability created a need for a new type of flight control system known as **fly-by-wire**. This approach implies that particular system components are interconnected electrically rather than mechanically and that they operate under the supervision of a computer responsible for monitoring, controlling, and coordinating the tasks. The fly-by-wire principle allows for the implementation of totally digital and highly redundant control systems reaching a remarkable level of reliability and performance [69]. The problem presented here is adapted from AP13.2 in *Modern Control Systems*.

Operational characteristics of a flight control system depend on the **dynamic stiffness** of an actuator, which represents its ability to maintain the position of the control surface in spite of the disturbing effects of random external forces. One flight actuator system consists of a special type of dc motor, driven by a power amplifier, which drives a hydraulic pump that is connected to either side of a hydraulic cylinder. The piston of the hydraulic cylinder is directly connected to a control surface of an aircraft through some appropriate mechanical linkage, as shown in Figure 13.2.

The plant model is given by

$$G_p(s) = \frac{1}{s(s + 1)} .$$

(13.1)

The **zero-order hold** is modeled by

$$G_o(s) = \frac{1 - e^{-sT}}{s} .$$

(13.2)

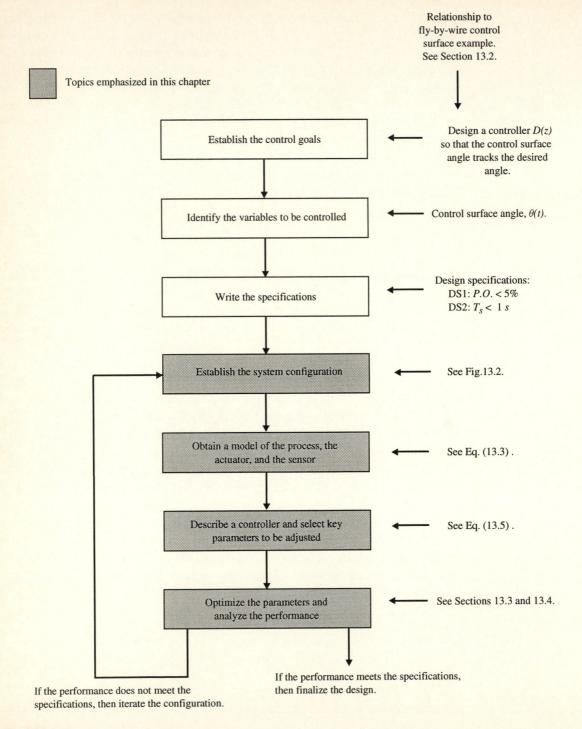

FIGURE 13.1
Elements of the control system design process emphasized in this chapter.

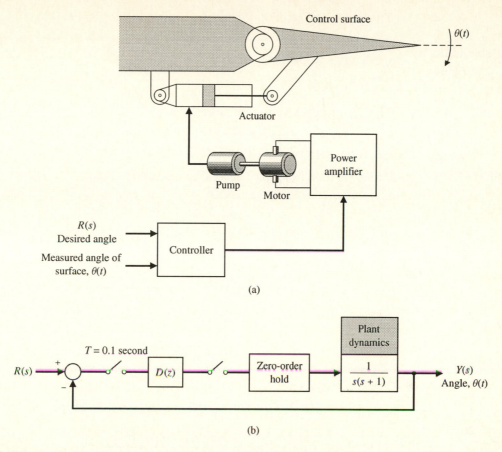

FIGURE 13.2
(a) Fly-by-wire aircraft control surface system and (b) block diagram. The sampling period is 0.1 second.

Combining the plant and the zero-order hold in series yields

$$G(s) = G_o(s)G_p(s) = \frac{1 - e^{-sT}}{s^2(s + 1)} . \qquad (13.3)$$

The control goal is to design a compensator, $D(z)$, so that the control surface angle $Y(s) = \theta(s)$ tracks the desired angle, denoted by $R(s)$. We state the control goal as

Control Goal   Design a controller, $D(z)$, so that the control surface angle tracks the desired angle.

The variable to be controlled is the control surface angle, $\theta(t)$:

Control surface angle, $\theta(t)$.

The design specifications are as follows:

DS1 Percent overshoot less than 5% to a unit step input.

DS2 Settling time less than 1 second to a unit step input.

We begin the design process by determining $G(z)$ from $G(s)$. Expanding $G(s)$ in Eq. (13.3) in partial fractions yields

$$G(s) = (1 - e^{-sT}) \left(\frac{1}{s^2} - \frac{1}{s} + \frac{1}{s+1} \right) ,$$

and

$$G(z) = \mathcal{Z}\{G(s)\} = \frac{ze^{-T} - z + Tz + 1 - e^{-T} - Te^{-T}}{(z-1)(z-e^{-T})} ,$$

where $\mathcal{Z}\{\cdot\}$ represents the **z-transform**. Since $T = 0.1$, we have

$$G(z) = \frac{0.004837z + 0.004679}{(z-1)(z-0.9048)} . \tag{13.4}$$

For a simple compensator, $D(z) = K$, the root locus is shown in Figure 13.3. For stability we require $K < 21$. Note that the stability region for discrete-time systems is inside the unit circle in the complex plane. Recall that for continuous-time systems, the stability region is the left half-plane.

Using an iterative approach (with MATLAB), we discover that as $K \rightarrow 21$, the step response is very oscillatory, and the percent overshoot is too large; conversely, as K gets smaller, the settling time gets too long, although the percent overshoot decreases. In any case the design specifications cannot be satisfied with a simple proportional controller, $D(z) = K$. We need to utilize a more sophisticated controller.

We have the freedom to select the controller type. As with control design for continuous-time systems, the choice of compensator is always a challenge and problem-dependent. Here we choose a compensator with the general structure

$$D(z) = K\frac{z-a}{z-b} . \tag{13.5}$$

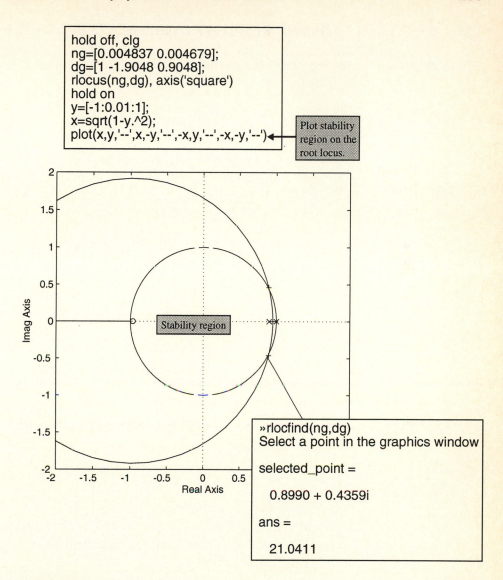

```
hold off, clg
ng=[0.004837 0.004679];
dg=[1 -1.9048 0.9048];
rlocus(ng,dg), axis('square')
hold on
y=[-1:0.01:1];
x=sqrt(1-y.^2);
plot(x,y,'--',x,-y,'--',-x,y,'--',-x,-y,'--')
```

Plot stability region on the root locus.

```
»rlocfind(ng,dg)
Select a point in the graphics window

selected_point =

   0.8990 + 0.4359i

ans =

   21.0411
```

Stability region

FIGURE 13.3
Root locus for $D(z) = K$.

Therefore, the key tuning parameters are the compensation parameters:

**Select Key Tuning
Parameters**

K, a, and b.

13.3 SETTLING TIME AND PERCENT OVERSHOOT SPECIFICATIONS

Settling Time For continuous systems we know that a design rule-of-thumb formula for settling time is

$$T_s \approx \frac{4}{\zeta \omega_n} ,$$

where we use a 2% bound to define settling. This design rule-of-thumb is valid for second-order systems with no zeros. So to meet the T_s requirement, we want

$$-\text{Re}(s_i) = \zeta \omega_n > \frac{4}{T_s} , \tag{13.6}$$

where s_i, $i = 1, 2$ are the dominant complex-conjugate poles. In the definition of the desired region of the z-plane for placing the dominant poles, we use the transform

$$z = e^{s_i T} = e^{-\zeta \omega_n \pm j \omega_n \sqrt{(1-\zeta^2)} T} = e^{-\zeta \omega_n T} e^{\pm j \omega_n T \sqrt{(1-\zeta^2)}} .$$

More discussion on the above transform can be found in Chapter 13 of *Modern Control Systems*. Computing the magnitude of z yields

$$r_o = |z| = e^{-\zeta \omega_n T} .$$

To meet the settling time specification, we need the z-plane poles to be inside the circle defined by

$$r_o = e^{-\frac{4T}{T_s}} , \tag{13.7}$$

where we have used the result in Eq. (13.6).

Consider the settling time requirement $T_s < 1$ second. In our case $T = 0.1$ second. From Eq. (13.7) we determine that the dominant z-plane poles should lie inside the circle defined by

$$r_o = e^{-\frac{0.4}{1}} = 0.67 .$$

Percent Overshoot As shown in Chapter 13 of *Modern Control Systems*, we can draw lines of constant ζ on the z-plane. The lines of constant ζ on the s-plane are radial lines with

$$\sigma = -\omega \tan(\sin^{-1} \zeta) = -\frac{\zeta}{\sqrt{1 - \zeta^2}} \omega .$$

Then, with $s = \sigma + j\omega$ and using the transform $z = e^{sT}$, we have

$$z = e^{-\sigma \omega T} e^{j \omega T} . \tag{13.8}$$

For a given ζ, we can plot Re$\{z\}$ vs Im$\{z\}$ for z given in Eq. 13.8 (see Figure 13.29 in *Modern Control Systems*).

If we were working with a second-order transfer function in the s-domain, we would need to have the damping ratio associated with the dominant roots be greater than $\zeta \geq 0.69$. When $\zeta \geq 0.69$, the percent overshoot for a second-order system (with no zeros) will be less than 5%. The curves of constant ζ on the z-plane will define the region in the z-plane where we need to place the dominant z-plane poles to meet the percent overshoot specification.

13.4   CONTROLLER DESIGN

The root locus in Figure 13.3 is repeated in Figure 13.4 with the stability and desired performance regions included. We can see that the root locus does not lie in the intersection of the stability and performance regions. The question is how to select the controller parameters K, a, and b so that the root locus lies in the desired regions.

One approach to the design is to choose a such that the pole of $G(z)$ at $z = 0.9048$ is cancelled. Then we must select b so that the root locus lies in the desired region. For example, when

$$a = -0.9048 \text{ and } b = 0.25 ,$$

the compensated root locus appears as shown in Figure 13.5. The root locus lies inside the performance region, as desired.

Using the rlocfind function, we can obtain a valid value of K. One valid value is

$$K = 70 .$$

Thus the compensator is

$$D(z) = 70\frac{s - 0.9048}{s + 0.25} .$$

The closed-loop step response is shown in Figure 13.6. Notice that the percent overshoot specification ($P.O. \leq 5\%$) is satisfied, and the system response settles in less than 10 samples (10 samples = 1 second since the sampling time is 0.1 second).

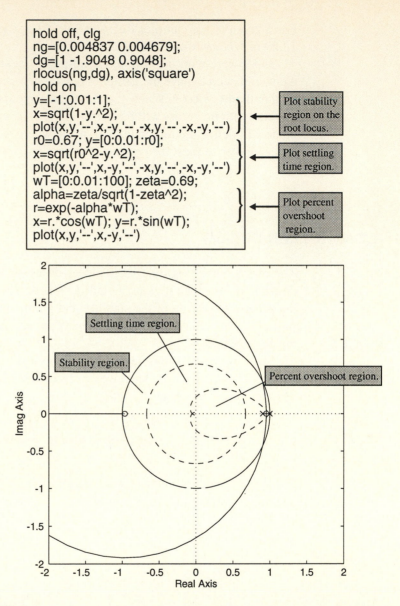

FIGURE 13.4
Root locus for $D(z) = K$ with the stability and performance regions shown.

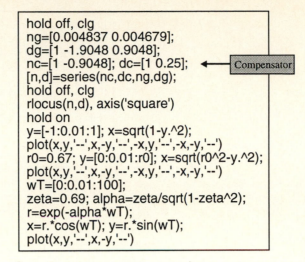

```
hold off, clg
ng=[0.004837 0.004679];
dg=[1 -1.9048 0.9048];
nc=[1 -0.9048]; dc=[1 0.25];     ← Compensator
[n,d]=series(nc,dc,ng,dg);
hold off, clg
rlocus(n,d), axis('square')
hold on
y=[-1:0.01:1]; x=sqrt(1-y.^2);
plot(x,y,'--',x,-y,'--',-x,y,'--',-x,-y,'--')
r0=0.67; y=[0:0.01:r0]; x=sqrt(r0^2-y.^2);
plot(x,y,'--',x,-y,'--',-x,y,'--',-x,-y,'--')
wT=[0:0.01:100];
zeta=0.69; alpha=zeta/sqrt(1-zeta^2);
r=exp(-alpha*wT);
x=r.*cos(wT); y=r.*sin(wT);
plot(x,y,'--',x,-y,'--')
```

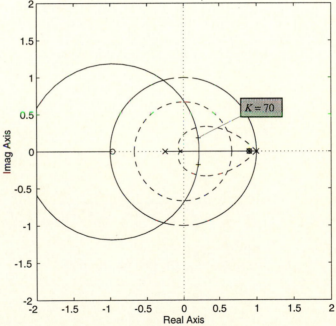

FIGURE 13.5
Compensated root locus.

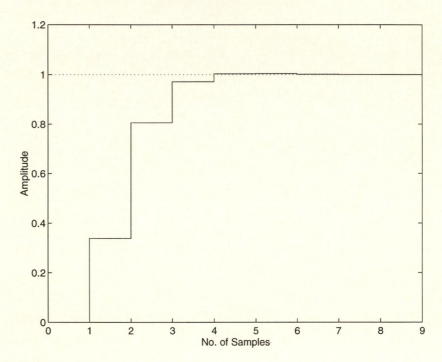

```
hold off, clg
ng=[0.004837 0.004679];
dg=[1 -1.9048 0.9048];
nc=70*[1 -0.9048];
dc=[1 0.25];
[n,d]=series(nc,dc,ng,dg);
[num,den]=cloop(n,d);
dstep(num,den,10)
```

FIGURE 13.6
Closed-loop system step response.

13.5 SUMMARY

In this chapter we considered the design of a digital control system. We used the fly-by-wire aircraft control surface problem to illustrate the root locus design method for discrete systems. Similar to the s-plane design approach, the z-plane root locus approach uses the concept of placing the dominant poles in the desired z-plane region defined by the performance specifications. The desired regions in the z-plane defined by the settling time and percent overshoot were described and placed on the root locus plot.

EXERCISES

E13.1   Design a digital control system to meet the following specifications:

**Design
Specifications**

DS1 Percent overshoot less than 10% to a unit step input.

DS2 Settling time less than 5 seconds to a unit step input.

Sketch the region in the z-plane where the dominant poles should be located to meet the design specifications. The sampling time is 0.25 seconds.

E13.2   For the fly-by-wire control surface problem of Section 13.2, design a digital controller to meet the following specifications:

**Design
Specifications**

DS1 Percent overshoot less than 2% to a unit step input.

DS2 Settling time less than 0.5 seconds to a unit step input.

E13.3   The fly-by-wire control surface problem in Section 13.2 has been re-engineered to meet tougher design specifications. In particular, the sampling time is now only 0.01 seconds. Design a digital controller to meet the following specifications:

**Design
Specifications**

DS1 Percent overshoot less than 1% to a unit step input.

DS2 Settling time less than 0.5 seconds to a unit step input.

Useful Design Formulas

The design formulas presented here are valid for second-order systems of the form

$$T(s) = \frac{\omega_n^2}{s^2 + 2\zeta\omega_n s + \omega_n^2} \, .$$

In some instances the formulas are approximations, even for second-order systems. Use the formulas with the knowledge that they may be successfully used in the beginning stages of the design process for systems of higher-order and systems with zeros. In most cases, some iterations of the design process will be required to fine-tune the performance to insure that all design specifications have been satisfied.

- Settling time (to within 2% of the final value)

$$T_s = \frac{4}{\zeta\omega_n}$$

- Percent overshoot

$$P.O. = 100e^{\frac{-\pi\zeta}{\sqrt{1-\zeta^2}}}$$

- Time-to-peak

$$T_p = \frac{\pi}{\omega_n\sqrt{1-\zeta^2}}$$

- Rise time (time to rise from 10% to 90% of final value)

$$T_{r_1} = \frac{2.16\zeta + 0.60}{\omega_n} \qquad (0.3 \leq \zeta \leq 0.8)$$

- Phase margin

$$P.M. = 100\zeta \qquad (\zeta \leq 0.7)$$

- Maximum magnitude of closed-loop frequency response

$$M_{p_\omega} = \frac{1}{2\zeta\sqrt{1-\zeta^2}} \qquad (\zeta \leq 0.7)$$

- Resonant frequency

$$\omega_r = \omega_n\sqrt{1 - 2\zeta^2} \qquad (\zeta \leq 0.7)$$

- Bandwidth

$$\omega_B = (-1.1961\zeta + 1.8508)\omega_n \qquad (0.3 \leq \zeta \leq 0.8)$$

REFERENCES

Chapter 1

1. M. M. Nakano and R. L. Willms, "Space Shuttle On-Orbit Flight Control Systems," *AIAA Guidance, Navigation, and Control Conference Proceedings*, Paper No. 82-1579, 1982, pp. 429–436.

2. G. T. Pope, "America's Last and Best Shuttle," *Popular Mechanics* June 1992, pp. 44–47.

3. G. Norris, "Boeing's Seventh Wonder," *IEEE Spectrum*, Vol. 32, No. 10, October 1995, pp. 20–23.

4. D. Hughes, "Fly-By-Wire 777 Keeps Traditional Cockpit," *Aviation Week & Space Technology*, McGraw-Hill Publication, May 1, 1995, pp. 42–48.

5. R. Reck, "Design Engineering," *Aerospace America*, American Institute of Aeronautics and Astronautics, Vol. 32, No. 12, December 1994, p. 74.

6. J. Binder, "Rapid Prototyping Gives Productivity a Boost," *Aerospace America*, American Institute of Aeronautics and Astronautics, Vol. 33, No. 10, October 1995, pp. 15–17.

7. MATLAB *User's Guide*, The MathWorks, Inc., Natick, Massachusetts, 1993.

8. SIMULINK *User's Guide*, The MathWorks, Inc., Natick, Massachusetts, March 1992.

Chapter 2

9. C. Nelson Dorny, *Understanding Dynamic Systems: Approaches to Modeling, Analysis, and Design*, Prentice-Hall, Englewood Cliffs, New Jersey, 1993.

10. T. D. Burton, *Introduction to Dynamic Systems Analysis*, McGraw-Hill, Inc., New York, 1994.

11. Katsuhiko Ogata, *System Dynamics*, Prentice-Hall, Englewood Cliffs, New Jersey, 1992.

12. J. D. Anderson, *Fundamentals of Aerodynamics*, McGraw-Hill, Inc., New York, 1991.

13. G. Emanuel, *Gasdynamics Theory and Applications*, AIAA Education Series, New York, 1986.

14. Arnold M. Kuethe and Chuen-Yen Chow, *Foundations of Aerodynamics: Bases of Aerodynamic Design*, 4th Edition, John Wiley & Sons, New York, 1986.

Chapter 3

15. M. Kaplan, *Modern Spacecraft Dynamics and Control*, John Wiley and Sons, New York, 1976.

16. J. Wertz, Ed., *Spacecraft Attitude Determination and Control*, Kluwer Academic Publishers, Dordrecht, The Netherlands, 1978 (reprinted in 1990).

17. W. E. Wiesel, *Spaceflight Dynamics*, McGraw-Hill, New York, 1989.

18. B. Wie, K. W. Byun, V. W. Warren, D. Geller, D. Long and J. Sunkel, "New Approach to Attitude/Momentum Control for the Space Station," *AIAA Journal Guidance, Control, and Dynamics*, Vol. 12, No. 5, 1989, pp. 714–722.

19. L. R. Bishop, R. H. Bishop, and K. L. Lindsay, "Proposed CMG Momentum Management Scheme for Space Station," *AIAA Guidance Navigation and Controls Conference Proceedings*, Vol. 2, No. 87-2528, 1987, pp. 1229–1236.

20. H. H. Woo, H. D. Morgan, and E. T. Falangas, "Momentum Management and Attitude Control Design for a Space Station," *AIAA Journal of Guidance, Control, and Dynamics*, Vol. 11, No. 1, 1988, pp. 19–25.

21. J. W. Sunkel and L. S. Shieh, "An Optimal Momentum Management Controller for the Space Station," *AIAA Journal of Guidance, Control, and Dynamics*, Vol. 13, No. 4, 1990, pp. 659–668.

22. V. W. Warren, B. Wie, and D. Geller, "Periodic-Disturbance Accommodating Control of the Space Station," *AIAA Journal of Guidance, Control, and Dynamics*, Vol. 13, No. 6, 1990, pp. 984–992.

23. B. Wie, A. Hu, and R. Singh, "Multi-Body Interaction Effects on Space Station Attitude Control and Momentum Management," *AIAA Journal of Guidance, Control, and Dynamics*, Vol. 13, No. 6, 1990, pp. 993–999.

24. J. W. Sunkel and L. S. Shieh, "Multistage Design of an Optimal Momentum Management Controller for the Space Station," *AIAA Journal of Guidance, Control, and Dynamics*, Vol. 14, No. 3, 1991, pp. 492–502.

25. K. W. Byun, B. Wie, D. Geller, and J. Sunkel, "Robust H_∞ Control Design for the Space Station with Structured Parameter Uncertainty," *AIAA Journal of Guidance, Control, and Dynamics*, Vol. 14, No. 6, 1991, pp. 1115–1122.

26. E. Elgersma, G. Stein, M. Jackson, and J. Yeichner, "Robust Controllers for Space Station Momentum Management," *IEEE Control Systems Magazine*, Vol. 12, No. 2, October 1992, pp. 14–22.

27. G. J. Balas, A. K. Packard, and J. T. Harduvel, "Application of μ-Synthesis Technique to Momentum Management and Attitude Control of the Space Station," *Proceedings of 1991 AIAA Guidance, Navigation, and Control Conference*, New Orleans, Louisiana, pp. 565–575.

28. I. Rhee and J. L. Speyer, "Robust Momentum Management and Attitude Control System for the Space Station," *AIAA Journal of Guidance, Control, and Dynamics*, Vol. 15, No. 2, 1992, pp. 342–351.

29. T. F. Burns and H. Flashner, "Adaptive Control Applied to Momentum Unloading Using the Low Earth Orbital Environment," *AIAA Journal of Guidance, Control, and Dynamics*, Vol. 15, No. 2, 1992, pp. 325–333.

30. X. M. Zhao, L. S. Shieh, J. W. Sunkel, and Z. Z. Yuan, "Self-Tuning Control of Attitude and Momentum Management for the Space Station," *AIAA Journal of Guidance, Control, and Dynamics*, Vol. 15, No. 1, 1992, pp. 17–27.

31. A. G. Parlos and J. W. Sunkel, "Adaptive Attitude Control and Momentum Management for Large-Angle Spacecraft Maneuvers," *AIAA Journal of Guidance, Control, and Dynamics*, Vol. 15, No. 4, 1992, pp. 1018–1028.

32. R. H. Bishop, S. J. Paynter, and J. W. Sunkel, "Adaptive Control of Space Station with Control Moment Gyros," *IEEE Control Systems Magazine*, Vol. 12, No. 2, October 1992, pp. 23–28.

33. S. R. Vadali and H. S. Oh, "Space Station Attitude Control and Momentum Management: A Nonlinear Look," *AIAA Journal of Guidance, Control, and Dynamics*, Vol. 15, No. 3, 1992, pp. 577–586.

34. S. N. Singh and T. C. Bossart, "Feedback Linearization and Nonlinear Ultimate Boundedness Control of the Space Station Using CMG," *AIAA Guidance Navigation and Controls Conference Proceedings*, Vol. 1, No. 90-3354-CP, 1990, pp. 369–376.

35. S. N. Singh and T. C. Bossart, "Invertibility of Map, Zero Dynamics and Nonlinear Control of Space Station," *AIAA Guidance Navigation and Controls Conference Proceedings*, Vol. 1, No. 91-2663-CP, 1991, pp. 576–584.

36. S. N. Singh and A. Iyer, "Nonlinear Regulation of Space Station: A Geometric Approach," *AIAA Journal of Guidance, Control, and Dynamics*, Vol. 17, No. 2, 1994, pp. 242–249.

37. J. J. Sheen and R. H. Bishop, "Spacecraft Nonlinear Control," *The Journal of Astronautical Sciences*, Vol. 42, No. 3, 1994, pp. 361–377.

38. J. Dzielski, E. Bergmann, J. Paradiso, D. Rowell, and D. Wormley, "Approach to Control Moment Gyroscope Steering Using Feedback Linearization," *AIAA Journal of Guidance, Control, and Dynamics*, Vol. 14, No. 1, 1991, pp. 96–106.

39. J. J. Sheen and R. H. Bishop, "Adaptive Nonlinear Control of Spacecraft," *The Journal of Astronautical Sciences*, Vol. 42, No. 4, 1994, pp. 451–472.

40. S. N. Singh and T. C. Bossart, "Exact Feedback Linearization and Control of Space Station Using CMG," *IEEE Transactions on Automatic Control*, Vol. Ac-38, No. 1, 1993, pp. 184–187.

Chapter 4

41. J. M. Maciejowski, *Multivariable Feedback Design*, Addison-Wesley, Wokingham, England, 1989.

42. *The American Medical Association Home Medical Encyclopedia*, Volume 1, Random House, New York, 1989, pp. 104–106.

43. J. B. Slate, L. C. Sheppard, V. C. Rideout, and E. H. Blackstone, "Closed-loop Nitroprusside Infusion: Modeling and Control Theory for Clinical Applications," *Proceedings IEEE International Symposium on Circuits and Systems*, 1980, pp. 482–488.

44. B. C. McInnis and L. Z. Deng, "Automatic Control of Blood Pressures with Multiple Drug Inputs," *Annals of Biomedical Engineering*, Vol. 13, 1985, pp. 217–225.

45. R. Meier, J. Nieuwland, A. M. Zbinden, and S. S. Hacisalihzade, "Fuzzy Logic Control of Blood Pressure During Anesthesia," *IEEE Control Systems*, December 1992, pp. 12–17.

46. L. C. Sheppard, "Computer Control of the Infusion of Vasoactive Drugs," *Proceedings IEEE International Symposium on Circuits and Systems*, 1980, pp. 469–473.

47. R. Vishnoi and R. J. Roy, "Adaptive Control of Closed-Circuit Anesthesia," *IEEE Transactions on Biomedical Engineering*, Vol. 38, No. 1, 1991, pp. 39–47.

Chapter 5

48. B. L. Stevens and F. L. Lewis, *Aircraft Control and Simulation*, John Wiley & Sons, New York, 1992.

49. B. Etkin and L. D. Reid, *Dynamics of Flight, 3rd Edition*, John Wiley & Sons, New York, 1996.

50. G. E. Cooper and R. P. Harper, Jr., "The Use of Pilot Rating in the Evaluation of Aircraft Handling Qualities," *NASA TN D-5153*, 1969.

51. USAF, "Flying Qualities of Piloted Vehicles," *USAF Spec.* MIL-F-8785C, 1980.

Chapter 6

52. L. Hatvani, "Adaptive Control: Stabilization," *Applied Control*, edited by Spyros G. Tzafestas, Marcel Decker, New York, 1993, pp. 273–287.

Chapter 7

53. *Space Station Reference Configuration Description*, Systems Engineering and Integration, Space Station Program Office, NASA Johnson Space Center, JSC-19989, August 1984.

54. P. Varaiya, "Smart Cars on Smart Roads," *IEEE Transactions on Automatic Control*, February 1993, pp. 195–207.

55. J. G. Kassakian, H.-C. Wolf, J. M. Miller, and C. J. Hurton, "Automotive Electrical Systems Circa 2005," Vol. 33, No. 8, *IEEE Spectrum*, August 1996, pp. 22–27.

Chapter 8

56. D. L. Smith, *Introduction to Dynamic System Modeling for Design*, Prentice-Hall, New Jersey, 1994.

57. D. Leonard, "Ambler Ramblin," Vol. 2, No. 7, *Ad Astra*, July-August 1990, pp. 7–9.

Chapter 9

58. J. Pretlove, "Stereo Vision," *Industrial Robot*, Vol. 21, No. 2, 1994, pp. 24–31.

59. M. W. Spong and M. Vidyasagar, *Robot Dynamics and Control*, John Wiley & Sons, New York, 1989.

60. R. C. Dorf, *Encyclopedia of Robotics*, John Wiley & Sons, New York, 1988.

Chapter 10

61. R. C. Dorf and A. Kusiak, *Handbook of Design Manufacturing and Automation*, John Wiley & Sons, New York, 1994.

62. A. G. Ulsoy, "Control of Machining Processes," *Journal of Dynamic Systems*, June 1993, pp. 301–310.

63. B. K. Bose, *Modern Power Electronics*, IEEE Press, New York, 1992.

Chapter 11

64. M. W. Spong and M. Vidyasagar, *Robot Dynamics and Control*, John Wiley & Sons, New York, 1989.

65. D. Luenberger, "Observers for Multivariable Systems," *IEEE Transactions on Automatic Control*, 11, 1966, pp. 190–197.

66. W. S. Levine, *The Control Handbook*, CRC Press and IEEE Press, Boca Raton, Florida, 1996.

Chapter 12

67. J. C. Doyle, A. B. Francis, and A. R. Tannenbaum, *Feedback Control Theory*, MacMillan, New York, 1992.

68. R. C. Dorf, *Electrical Engineering Handbook*, CRC Press, Boca Raton, Florida, 1993.

Chapter 13

69. V. Skormin, "On-line Diagnostics of a Self-Contained Flight Actuator," *IEEE Transactions on Aerospace and Electronic Systems*, January 1994, pp. 130–141.

acker, 201
Ackermann's formula, 201
advantages of feedback, 62
airplane lateral dynamics, 86
approximating transfer functions, 91
asymptotes, 134
atmospheric drag, 50

bandwidth, 148
Bernoulli's equation, 22
BIBO stability, 105
bode, 162
bode plot, 140
Bode plot, 162
break frequency, 173, 175

characteristic roots, 195
complementary sensitivity function, 65
control goal
    airplane lateral dynamics, 86
    automobile velocity control, 131
    blood pressure control, 72
    diesel electric locomotive, 204
    fly-by-wire control system, 233, 234
    hot ingot robot control, 160
    international space station, 50
    milling machine, 186
    robot-controlled motorcycle, 112
    six-legged ambler, 146
controllability, 199
controllability matrix, 199

convolution integral, 108
Cooper-Harper pilot opinion ratings, 88
ctrb, 204

derivative controller, 129
design gaps, 4
design goal
    digital audio tape, 223
design process, 5
design specifications
    airplane lateral dynamics, 88
    automobile velocity control, 132
    blood pressure control, 73
    diesel electric locomotive, 206
    digital audio tape, 223
    hot ingot robot control, 160
    milling machine, 187
    robot-controlled motorcycle, 112
    six-legged ambler, 147
digital compensator design
    fly-by-wire control system, 237
disturbance rejection, 68
Dutch roll, 89

eigenvalues, 195
error signal, 62
Euler identity, 165

fluid flow system, 18
frequency response, 145

gain margin, 158, 167
gravity force, 50

hand sketching
    Bode plots, 140
    root locus, 122

impulse response data
    blood pressure control, 75
    milling machine, 185
incompressible fluid, 20
integral controller, 129
international space station, 43
    aerodynamic torque, 52
    body coordinate system, 48
    control moment gyros, 45
    gravity gradient torque, 46, 51
    inertial coordinate system, 46
    local vertical/local horizontal coordinate system, 47
    spherical coordinate system, 46
    torque equilibrium attitude, 54
irrotational flow, 21

lag compensator, 181
    milling machine, 187
lead compensator, 181
linearization
    fluid flow, 23
loop gain, 64

margin, 162, 167
mathematical model, 17
MATLAB analysis
    automobile velocity control, 135
    blood pressure control, 78

diesel electric locomotive, 210
digital audio tape, 224
fluid flow, 28
fly-by-wire control system, 237
hot ingot robot control, 177
international space station, 56
milling machine, 190
RLC series circuit, 43
robot drive train dynamics, 198
robot-controlled motorcycle,
    117
rolling carts, 40
six-legged ambler, 151
measurement noise attenuation,
    70
minimal complexity, 72
multiplicative uncertainty
    representation, 219

Nichols diagram, 176
nonlinear models
    fluid flow, 23
    international space station, 53
Nyquist plot, 163
    time-delay, 171
Nyquist theorem, 163

observability, 203
observability matrix, 203
obsv, 204

pade, 170
Padé, 168
percent overshoot, 173
    fly-by-wire control system, 236
phase margin, 173
phase margin, 158
proportional controller, 129
proportional-derivative controller,
    126
proportional-integral controller,
    173
proportional-integral-derivative
    controller, 76, 126, 130
    digital audio tape, 224
    six-legged ambler, 147

rapid prototyping, 5
reduced sensitivity, 66
resonant frequency, 149
RLC series circuit, 40
rlocfind, 135
    digital audio tape, 224
    milling machine, 188
robustness, 216
roll subsidence, 89
rolling carts, 37
    free-body diagram, 37
root locus, 121
    discrete-time system, 234
Routh-Hurwitz method, 110

select key tuning parameters
    airplane lateral dynamics, 90
    blood pressure control, 76
    diesel electric locomotive, 206
    digital audio tape, 224
    fly-by-wire control system, 235
    milling machine, 187
    robot-controlled motorcycle,
        114
    six-legged ambler, 148
sensitivity analysis, 65
sensitivity function, 64
settling time
    automobile velocity control,
        133
    fly-by-wire control system, 236
    six-legged ambler, 149
simplified block diagram, 2
SIMULINK, 7
    File, 14
        Close, 14
        Save, 14
    Simulation, 12
        Parameters, 12
        Start, 12
        Stop, 14
    standard block library, 8
        Connections, 8
        Discrete, 8
        Extras, 8
        Linear, 8

Nonlinear, 8
Sinks, 8
Sources, 8
SIMULINK analysis
    airplane lateral dynamics
        simulation analysis, 100
        simulation development, 95
    diesel electric locomotive, 211
Skunk Works, 128
small gain theorem, 219
space shuttle, 2
    attitude hold control, 2
spiral mode, 89
ss2tf, 193
state feedback control law, 200
state feedback controller design
    diesel electric locomotive, 206
state variable model, 193
    international space station, 54
    RLC series circuit, 42
    robot drive train dynamics, 196
    rolling carts, 39
state variables, 193
steady flow, 21
steady-state tracking error
    diesel electric locomotive, 208
system stability
    mass-spring-damper, 105
system stability, 105

tf2ss, 193
time-delay, 138
transfer function from state
        variable model, 193
transfer function model
    airplane lateral dynamics, 89
    digital audio tape, 222
    fluid flow, 25
    fly-by-wire control system, 231
    hot ingot robot control, 158
    human patient, 75
    milling machine, 184
    robot-controlled motorcycle,
        112
    six-legged ambler, 146

uncertain time-delays, 217

variable to be controlled
   airplane lateral dynamics, 88
   automobile velocity control,
     131
   blood pressure control, 72

diesel electric locomotive, 205
digital audio tape, 223
fly-by-wire control system, 234
milling machine, 186
robot-controlled motorcycle,
   112
six-legged ambler, 146

viscosity, 21

Wright brothers, 85